THE CUTTING EDGE OF
TRIBOLOGY
A Decade of Progress in Friction, Lubrication, and Wear

THE CUTTING EDGE OF
TRIBOLOGY
A Decade of Progress in Friction, Lubrication, and Wear

Nicholas D Spencer
ETH Zürich, Switzerland

Wilfred T Tysoe
University of Wisconsin-Milwaukee, USA

W🌐 World Scientific

NEW JERSEY · LONDON · SINGAPORE · BEIJING · SHANGHAI · HONG KONG · TAIPEI · CHENNAI

Published by

World Scientific Publishing Co. Pte. Ltd.

5 Toh Tuck Link, Singapore 596224

USA office: 27 Warren Street, Suite 401-402, Hackensack, NJ 07601

UK office: 57 Shelton Street, Covent Garden, London WC2H 9HE

British Library Cataloguing-in-Publication Data
A catalogue record for this book is available from the British Library.

THE CUTTING EDGE OF TRIBOLOGY
A Decade of Progress in Friction, Lubrication and Wear

ISBN 978-981-4656-55-9
ISBN 978-981-4663-23-6 (pbk)

In-house Editor: Rhaimie Wahap

Typeset by Stallion Press
Email: enquiries@stallionpress.com

Dedication

We dedicate this book to practicing tribologists worldwide.

Preface

In 2003, the Society of Tribologists and Lubrication Engineers was planning to update their image by changing the format and title of their society magazine, then called Lubrication Engineering. The aim was to capture the breadth of the society and the new magazine, launched in late 2003, was entitled Tribology and Lubrication Technology (TLT). A decade earlier, we had taken the step, at the behest of Gabor Somorjai, to launch a new journal in the area of tribology, named Tribology Letters, based on the successful format of Catalysis Letters, launched some years earlier by Somorjai and John Thomas. The aim of Tribology Letters was to be a leader in the area of tribological science, rather than having the engineering emphasis that characterized most tribology journals at that time. After a brief visit to Berkeley to visit Gabor Somorjai, to pick up hints on how to launch a journal, we officially launched Tribology Letters in mid-1995. We realized that it would be helpful to counterbalance our emphasis on tribological science by having a sister journal that emphasized the more engineering aspects of tribology, such as Tribology Transactions. We therefore entered into successful negotiations with STLE for Tribology Letters to become part of their official journal family. Consequently, it was rather natural that STLE contacted us when they were thinking about creating a new column in TLT that would keep their readers abreast of new developments in tribology. We agreed that we would jointly write a bimonthly column. While we have tried to include articles on the most exciting aspects of tribology that came to our attention, there has been an inevitable bias towards articles that have appeared in Tribology Letters.

Neither of us thought that we would be able to keep writing articles for very long — the challenge of writing to a deadline is completely different from scientific writing, where publication is only considered when the data have been obtained, analyzed, and hopefully understood. But we were wrong, and are still cheerfully turning out columns in our second decade. Another challenge was to come up with a name and, after some discussion we settled on (perhaps the rather corny) Cutting Edge. This set the stage for our future titles, which, were often a play on old jokes (Karma Runs Over Dogma), an attempt at new ones (Alcohol Gets You no Wear) or, at the very least, alliterative (The Continuing Contact Conundrum).

When the 10th anniversary of the Cutting Edge arrived, and with the 20th anniversary of Tribology Letters on the horizon, we approached World Scientific to see if they would be interested in putting together this anthology of past articles as part of TLT's anniversary celebrations, and they kindly agreed.

In 2013 we found ourselves at a number of the same conferences, in various parts of the world. This was a good opportunity to start to assemble this collection. Discussing how to do this on a balcony in Sardinia, we found that the subjects we had chosen over the years for the various articles fell quite naturally into distinct categories, which now form the basis for each chapter. These include our musings on Biotribology, Friction Fundamentals, Hard-Drive Lubrication, New Materials and Methods, Opinions and People, The Contact Conundrum, Tribochemistry and Weird and Wonderful Effects.

We would like to thank Karl Phipps and Tom Astrene at STLE for working with us on the production of these columns and now and then tolerating our missed deadlines, and we hope that you enjoy browsing through the last decades of our thoughts on what is exciting about our field. Many of these columns were written at the weekends, and therefore we also acknowledge the patience of our wives, Cristina Tysoe and Jennifer Davidson, over the last ten years. Finally, we would like to thank Josephine Baer for her great help during the preparation of the manuscript.

Nic Spencer, Zollikon, Switzerland
Eddy Tysoe, Milwaukee, USA

Contents

Opinions and People

Sometimes we have taken the opportunity to express opinions on topics we consider significant for our field. These have ranged from reports on workshops or conferences that we viewed as important, developments in tribology that could have an impact on the way people do or report research, or our current views on the state of the field.

Our "op-ed" pieces over the last decade included our initial column on the nature of tribological science (*Laws of Past Slowly Foster Today's Technology,* October, 2003), the proposed introduction of a new unit for wear, the *Bowden* (*Some Wear or no Wear*, December, 2005), the National Science Foundation's workshop on Grand Challenges in Tribology (*Frontiers of Tribological Research*, June, 2005), and reports on two important international meetings on tribology fundamentals: the Trends in Nanotribology conference in Trieste, Italy in 2010 (*The Fundamentals of Friction*, December 2010) and the first Faraday Meeting on Tribology, held in Southampton, UK, in 2012 (*Advancing our Understanding,* August 2012).

As editors-in-chief of Tribology Letters, we have also highlighted new aspects of our journal, such as the inclusion of a new type of article, aimed at establishing "best practices" in experimental tribological research (*Doing it Right*, April 2009), and our focus issue on the way in which experiment and theory can now be combined into collaborative research, to the benefit of a deeper understanding of tribological phenomena (*Experiment and Theory — Rubbing Along Together*, June 2013).

We have also felt the need to pay homage to great tribologists, upon their passing. Over the history of our column, these have included David

Tabor (*Remembering David Tabor*, August, 2006), Mike Gardos (*Tribute to the late, great Mike Gardos*, June, 2004), and our combined tribute to Sanjay Biswas and Brian Briscoe (*2013 Sees the Passing of Two Eminent Tribologists*, February, 2014). All of these influential scientists were also editorial board members of Tribology Letters.

Laws of Past Slowly Foster
Today's Technology

Developments in the late 17th Century laid the basis for what we understand as tribology today

The title of this column, "Cutting Edge," was chosen with a very definite purpose. Every other month we'll bring you news about the most recent advances in tribology research and discuss how they are relevant to you. For our inaugural issue, however, we've chosen to look backward for a moment and cast the spotlight on two scientific giants. For as all scientists know, the roads we travel in the future were built, stone by stone, in the past.

The latter half of the 17th Century was a remarkable period in the development of our modern view of the scientific method. It was also the time when two momentous works dealing with the motion of bodies were published. In 1687 Isaac Newton published *Philosofiæ Naturalis Principia Mathematica* ("The Mathematical Principles of Natural Philosophy"). The treatise examined the motion of non-contacting bodies. In 1699 Guillaume Amontons, following the work of Leonardo Da Vinci, published a paper in Paris in *Memoires de l'Académie des Sciences* entitled *De la Resistance Causéedans les Machines* ("On the Resistance Caused in Machines").

That paper established Amontons' First Law, which states that frictional force is proportional to the normal force between the bodies in contact. The ratio between the two is the friction coefficient, and Amontons believed this to be a universal constant with a value of 0.3.

Although initially propounded at almost the same time, the subsequent evolution of these laws has been completely different.

Sir Isaac Newton, 1643–1727.

Under most conditions, Newton's laws have proven remarkably precise, enabling us to project objects with great accuracy, including from the earth to the moon. Einstein's modifications to Newton's classical theory at velocities approaching the speed of light resulted in the theory of relativity. Applied to microscopic particles, where the very act of measuring them has to be incorporated into the theory, the modifications resulted in the development of quantum mechanics, which provided the theoretical basis for a vast array of modern technologies.

Guillaume Amontons (1663–1705). In addition to working on friction and gas laws, Amontons was also an inventor. Here he demonstrates his optical telegraphy system to the Dauphin of France. This demonstration is thought to have taken place sometime between 1688 and 1695. The system used a series of signalling stations with a wooden arm that could be moved into different positions to represent letters of the alphabet. The signals would be replicated in successive stations to pass a message by signalmen using telescopes to view the station before them in line. Amontons' system was never put into practice. From La télégraphie historique: depuis les temps les plus reculés jusqu'à nos jours, Alexis Belloc, (Paris: Fermin-Didot, 1888). Sheila Terry/Science Photo Library, by kind permission.

The evolution of Amontons' law, by contrast, has been much slower. Certainly, the pioneering work from the Tabor and Greenwood groups in Cambridge has taught us that the topography of contacting surfaces is central to understanding Amontons' law and that a relatively small proportion of the whole surface is actually in contact, with the contact area at the tips of asperities at the contacting interface indeed being proportional to the load. However, the way in which energy is dissipated during sliding is still not understood! The ability to predict friction reliably still eludes our grasp. Is this because the descendants of Amontons are less intelligent than those of Newton? Of course not! The truth, rather, is that the systems we tribologists deal with are far more complex, involving innumerable interactions between material, chemical and mechanical properties.

The technological developments of the last few decades since the pioneering work of Tabor — the ability to obtain well-characterized surfaces and measure their properties in ultrahigh vacuum, having exquisitely sensitive probes such as atomic force microscopes and the surface forces apparatus, as well as more powerful computers — have put an unprecedented array of tools into the hands of the modern tribologist. The developments in the next few years, in both experiment and theory, will undoubtedly allow the descendants of Amontons to catch up with those of Newton. It's an exciting world for those of us involved in the science of tribology.

Tribology and Lubrication Technology
October 2003, 59(10) p64

Some Wear or no Wear?

A seemingly innocent suggestion for a new wear unit, the Bowden, puts the cat among the pigeons in the normally reserved International Research Group on Wear of Engineering Materials

There's nothing like an animated discussion to enliven a field. The topics of wear and wear units (not generally cause for great emotion) have recently been brought to life by a controversial proposal originating from the International Research Group on Wear of Engineering Materials (IRG-OECD), which met in June (in 2005) Uppsala, Sweden. Kenneth Holmberg, who chairs the group and represents the majority opinion, points out that while the friction coefficient, μ, is a useful and widely used entity, no equivalent parameter exists for wear, leading to a disparate and confusing situation in wear reporting. He makes two main proposals, firstly that a minimum list of characteristic parameters that specify the contact situation in a wear experiment be drawn up, to facilitate the interpretation and reproduction of reported wear data. This list would include geometry and roughness, energy input, materials properties, environment, and characterization of the sliding surfaces and wear debris. The second proposal is that a new wear unit, the Bowden (B), be introduced, based purely on the ratio of the wear volume and (load $\times$ sliding distance), having the value 10^{-6} $mm^3N^{-1}m^{-1}$. The size of this unit is particularly convenient, since 1 B corresponds to normal wear, 100 B to high wear, and 0.01 B to low wear. Holmberg believes that standardization in the use of the Bowden would facilitate comparisons between studies carried out in different laboratories, and the minimum list of parameters would ultimately lead to more reliable wear reporting. While there are clear cases where alternative wear-reporting methods would be more appropriate, such as the use of wear depth, where tolerances are important, or the application of wear volume in the case of erosive wear, Holmberg estimates that 70–80% of all wear reporting could be based on the Bowden.

Fig. 1 Frank Philip Bowden (1903–1968), by kind permission from the Cavendish Laboratory, University of Cambridge.

A number of distinguished tribologists and journal editors have joined this discussion, and there is also broad support for an improvement in wear-reporting procedures and the minimum list of characteristic parameters. However, there is skepticism that a new unit, especially one that is not SI, will really help the situation. In general, a unit implies the acceptance of a physical relationship or definition. When we measure pressure in Pascals (Pa, Nm^{-2}), for example, we are defining pressure as force divided by area, and the pressure will remain the same, no matter what the area. In fact, we don't even have to think about the area, when we read the pressure on a gauge. In tribology, our "laws" are not physical laws, but are empirical observations that work for much of the time. In the case of friction, Amontons' Law works astonishingly well, and a simple definition of friction coefficient (friction force / applied load) is very useful. Nevertheless people do sometimes get into difficulties in cases where adhesive forces enter the picture, and the definition of a friction coefficient is less obvious — and less useful. In the case of wear, the definition of a Bowden as 10^{-6} $mm^3 N^{-1} m^{-1}$ carries with it the inherent implication that wear rates are load-independent, as well as independent of sliding distance. This is far from being always true, let alone a physical law! Another, more emotional objection that has been raised to the Bowden is the name. It is clear that Frank Philip Bowden (Figure 1), while a great tribological scientist, worked relatively little on wear problems. There are a number of other tribologists who one could think of honoring in this way, notably Archard, who is best known for his wear equation. Of course, a drawback to this proposal is that the abbreviation would be "A" or "Ar", which have other meanings. A more unusual suggestion is the "wearon", in analogy with electron and photon, but here, again, the name carries with it the implication of a quantized phenomenon rather than a convenient index.

All of us should care about improving the quality of reported tribological data. The question is how best to do it without creating more confusion.

Tribology and Lubrication Technology
December 2005, 61(12) p64

Frontiers of Fundamental Tribological Research

A National-Science-Foundation-sponsored meeting discusses future research directions in tribology

While a large amount of tribological research and development is carried out by industry, the vast majority of research into the fundamentals of tribology in the United States is funded by the National Science Foundation. These projects are aimed at understanding tribology, often at a molecular level, and exploring new concepts and ideas.

Periodically, the National Science Foundation holds workshops where a number of the leaders in a particular field are brought together to define the outstanding challenges, and ultimately to assist in deciding how scarce resources should be allocated to address these challenges. Such a workshop was held on tribology late in 2004 in Houston, Texas.

The workshop recognized the remarkable advances that have been made over the last decades in our understanding of friction and wear and how the field has benefited from the participation of physicists, chemists and materials scientists to complement the more traditional engineering approaches. Indeed, it is interesting to note that the latest Tribology Gold Medal recipient, Professor Hugh Spikes of Imperial College, London is, by training, a chemist.

These advances have had a major impact on technology, some notable examples being hard-disk-drive technologies, where an exponential increase in storage capacity has been achieved by the development of protective carbon overcoats and effective lubricants, and micro-electromechanical devices (MEMS), which are currently being

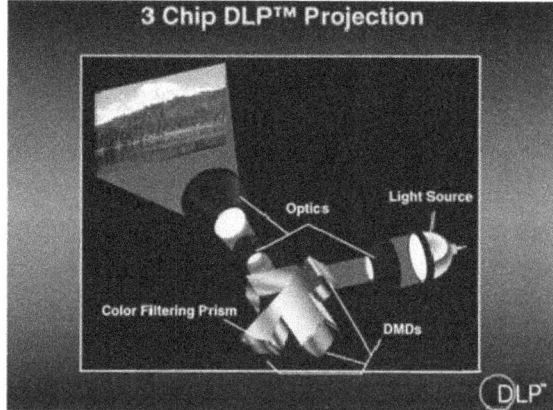

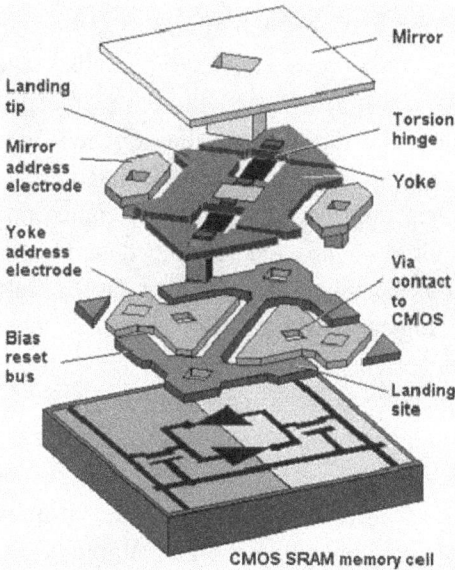

Fig. 1 Schematic depiction of the construction of the digital light processing (DLP) mirror and an illustration of its use in a projection system. By kind permission of Springer Science+Business Media from Reference [1].

used as image-projection devices based on digital light processing micromirrors that rapidly deflect light beams (Fig. 1). The successful development of this latter technology relied on finding tribological solutions to the problems of stiction and irreversible adhesion of the micromirrors, which had led to device failure.

Tribology in extreme environments was identified as an example of an important future technological challenge that is particularly critical as we contemplate moving to a hydrogen economy. The role of tribology in improving energy efficiency and security, and minimizing the impact to the environment, as well as biotribology for improving the lifetime of hip implants, are other emerging areas. Additional work still needs to be done to maintain the exponential growth in hard-disk-drive storage capacities and to solve ongoing problems in implementing MEMS technologies.

All of these challenges and many others will require that we improve our understanding of the tribological contact. A key to achieving this goal will require the development of new theoretical and experimental methods and the forging of closer links between experimentalists and theorists.

The development of scanning-probe microscopes has opened up a new world of "nanotribology" and improved our understanding of contact mechanics (Fig. 2). Further improvements, both in nano- and conventional tribometers are required to provide reliable data for the theorists to analyze. A wide range of analytical techniques is currently being applied to study the tribological interface, but many of these examine the surface after the event. Being able to monitor the interface *in situ* would significantly contribute to our understanding of the significant chemical and morphological changes at the interface and the way in which energy is dissipated there.

Molecular-dynamics simulations have been one of the primary approaches for investigating tribological phenomena on the atomic scale, but these are limited to small sizes and times and thus often miss important phenomena, such as wear or surface chemistry. Modifications of these approaches are required to provide a more realistic link to experiments, perhaps by linking atomic-scale approaches to continuum models or by carrying out quantum calculations.

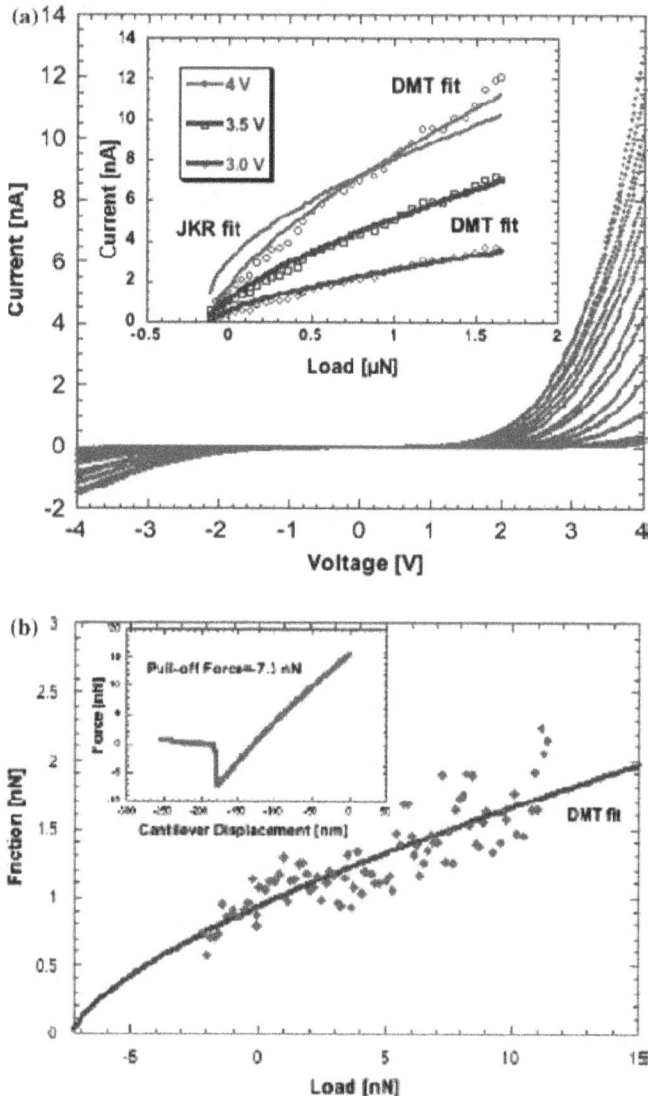

Fig. 2. (a) The contact area measured from the current between the tip and substrate showing that this is well described by DMT theory. Similarly, the friction measured using AFM also varies according to DMT theory, illustrating the friction being proportional to the contact area (b). By kind permission of Springer Science+Business Media from Reference [1].

Based on the progress that has been made over the past decade, it is clear that the field of tribology is poised to address many of the outstanding issues in the area, and to truly understand what is happening at a moving, contacting interface. In turn, these fundamental advances will lead to substantial technological improvements. One of the hallmarks of tribological research is the rapidity with which improvements in our fundamental understanding can be translated into technological advances. This is indeed an exciting time to be a tribologist.

Tribology and Lubrication Technology
June 2005, 61(6) p88

Further Reading:

[1] Perry, S.S., and Tysoe, W.T. (2005) Frontiers of Fundamental Tribological Research, Tribology Letters, 19, pp. 151–161.

The Fundamentals of Friction

Leonardo da Vinci's heirs explore the mechanisms behind friction and wear

As tribologists, we face big challenges. The world needs lower-friction bearings to reduce energy losses, lower-wear mechanical systems to lower maintenance costs, and new lubricant additives to lower environmental impact. Our community can help, but whatever kind of tribologist you are — scientist or engineer, bearing designer or molecular dynamicist — one thing should be clear: fundamental research is the lifeblood of the field.

Strangely enough, this is not a view that is universally shared by funding agencies, and academics often have an extraordinarily hard time to support their fundamental research activities. As often as not, fundamental research is "piggybacked" onto industrially supported, problem-solving projects, and therefore not always given the highest priority.

In Europe, there is a glimmer of hope that this situation may not be inevitable. On the continent where Leonardo da Vinci first observed the friction-load proportionality, later rediscovered and codified by Amontons, a collaborative initiative by the European Science Foundation (ESF), called Friction and Adhesion in Nanomechanical Systems (Fanas) is underway. In this context, the ESF recently sponsored a conference entitled "Trends in Nanotribology" at the International Centre for Theoretical Physics in Trieste, Italy. Selected papers from the meeting have recently appeared in a special issue of STLE-affiliated journal, *Tribology Letters* [1].

New tribological phenomena are the focus of research in the Fanas program, including areas such as superlubricity, wearless sliding, control of frictional properties, bridging the gap between the nano-, micro- and macroscales, manipulation of nanoparticles on surfaces, and aqueous lubrication. These are being addressed in collaborative projects taking place in many European countries and Israel, some with the involvement of researchers from the USA.

Fig. 1 Leonardo da Vinci's tribometer, from his notebook, ca. 1480.

Highlights of the Trieste meeting included both experimental and modeling studies, addressing atomic-level understanding of both friction and wear. In a contribution from Itay Barel and Michael Urbakh from Tel Aviv University, Israel, collaborating with Lars Jansen and André Schirmeisen at the University of Münster, the temperature dependence of friction was investigated [2]. The focus of the study involves the thermally activated detachment and reattachment of multiple contacts during sliding, which leads to a complex energy landscape, but one in which higher temperatures might be expected to lead to the surmounting of energy barriers, the breaking of contacts, and thus more facile slip and lower friction. Working against this trend is a thermally activated reattachment mechanism, which leads to higher friction at higher temperatures. The result is a peak in dry sliding friction at low temperatures (around 100 K), which the authors have demonstrated experimentally for a variety of different materials (silicon, silicon carbide, sodium chloride, graphite), and also modeled as a set of multiple contacts, with different activation energies for formation and rupture. The model was able to predict the experimental behavior with impressive accuracy, using very few fitting parameters, and thus initiating a new approach to understanding the dynamics of nanoscale dry sliding.

Thermal activation is also the central issue in a paper by Bernd Gotsmann and Mark Lantz, from IBM Research, Zurich, Switzerland, and Tevis Jacobs and Robert Carpick from the University of Pennsylvania [3]. In this case, atomic-scale wear is the focus of the investigation, and the authors ponder the extent to which transition-state theory — the chemist's standard quantitative approach to understanding the temperature dependence of chemical reactions — could be applied to this phenomenon. Their answer was that the approach can, indeed, explain atomic-scale wear phenomena, although caution is required in applying this model. Measurements of temperature-dependent wear in the atomic force microscope, which are quite scarce at present, are needed to identify relevant activated transitions, and to disentangle thermally activated and velocity-dependent effects on wear.

The collection of papers in *Tribology Letters* is merely a reflection of some of the research taking place within the Fanas program, which will involve further workshops, exchanges, and training courses over the next year.

Tribology and Lubrication Technology
December 2010, 66(12) p80

Further Reading:

[1] Gulseren, O., Manini, N., Meyer, E., Tosatti, E., Urbakh, M., and Vanossi, A. (2010). New Trends in Nanotribology, Tribology Letters, 39, p227.

[2] Barel, I., Urbakh, M., Jansen, L. and Schirmeisen, A. (2010). Macroscropic Temperature Dependence of Friction at the Nanoscale: When the Unexpected Turns Normal, Tribology Letters, 39, pp. 311–319.

[3] Jacobs, T.D.B., Gotsmann, B., Lantz, M.A., and Carpick, R.W. (2010). On the Application of Transition State Theory to Atomic-Scale Wear, Tribology Letters, 39, pp. 257–271.

Advancing Our Understanding

Discussions at a new symposium portray the depth and breadth of tribology research

The Discussions of the Faraday Society are a firm fixture of the British chemistry landscape. These meetings typically deal with the interfaces of physical chemistry to other fields, and have a unique format. Virtually all presented papers are written in advance and circulated to those planning to take part, who are expected to read them prior to the meeting. During the meeting itself, authors have five minutes to refresh the memories of those present, after which twenty minutes or so of discussion can take place. After the meeting, the papers, as well as transcripts of the ensuing discussions, are published by the Royal Society of Chemistry. Professor Rob Wood, of the National Centre for Advanced Tribology at Southampton (nCATS) took the bold step of organizing Faraday Discussions 156 — the first one on Tribology — and invited a host of well-known tribologists covering the areas of Biotribology, Predictive Modelling, Smart Surfaces, and Future Lubrication Systems. The quality of both talks and discussions was extremely high, and illustrated the value of questioners having time to read papers carefully and think in advance.

Some papers were syntheses of previous work, such as the very useful and comprehensive contribution of Dr Ian Taylor, Shell, UK, on Tribology and Energy Efficiency, in which he focussed on the impact of molecular structures of lubricant base oils and additives on friction reduction in machine elements and on overall fuel consumption in vehicles. Prof. Duncan Dowson, University of Leeds, UK, started the meeting off with a fascinating survey of the breadth of biotribology (for which, as Introductory Speaker, he was allowed a full half hour!).

The Faraday Discussions were named after Michael Faraday, an 18th Century British chemist, physicist and philosopher. (Photo: Millikan and Gale's Practical Physics (1922).)

Other authors dealt with modelling studies, carried out with many different computational methods. In one paper, Prof. Pwt Evans, of the University of Cardiff, UK, described new approaches to the prediction of fatigue failure in lubricated contacts. This work is of great importance to power-generation systems, such as wind turbines, where problems such as tooth breakage in gears are common. The approach involved constraining EHL calculations with contact mechanics, using measured surface roughness profiles, and could lead to the construction of a valuable predictive tool. Dr Werner Österle of the Federal Institute for Materials Research and Testing, in Berlin, Germany, described his use of the Movable Cellular Autonoma technique to understand the roles of different components in brake linings in determining both the friction and structural changes in the contacting surfaces. Interestingly, the formation of a nanocrystalline layer, as previously discussed in Cutting Edge [1] in the context of molecular dynamics simulations, was predicted to occur when those combinations of components were present that appeared to lead to smooth braking.

Fundamental experimental tribology research was also presented at Faraday Discussion 156. Prof. Liliane Léger, of the Université Paris-Sud, France described how pillar-structured silicone rubber surfaces lead to higher contact-area-normalized friction than smooth ones, suggesting that the elastic energy stored in the pillars is released after contact, leading to a more effective dissipation process. Professor Prof. Graham Leggett, of the University of Sheffield, UK, showed how hydrogen-bonding interactions between liquids and surfaces can dramatically affect tip-sample interactions in tribological experiments carried out in the atomic force microscope, raising important questions about published results on the nanotribology of self-assembled monolayers under liquids.

The papers (and discussions) from Faraday Discussions 156 on Tribology [2], make interesting reading, and are available from the Royal Society of Chemistry.

Tribology and Lubrication Technology
December 2012, 64(8) p56

Further Reading:

[1] Tysoe, W.T. and Spencer, N.D. (2008). Analyzing High-Speed Sliding, Tribology and Lubrication Technology, 64, pp. 56.

[2] Spencer, N.D. *et al.* (2012). Tribology, Faraday Discussions, 156, pp. 435*ff*

Doing It Right

A new section of Tribology Letters improves our ability to generate accurate measurements and share information

Editing a top tribology journal can be fun, but can have its downsides. Now and then we receive papers describing measurements that were simply not carried out properly, and are dominated by artifacts. One problem is that much of the know-how involved in tribological measurements is passed from generation to generation of graduate students, rather than being readily available in books. Consequently, newly established laboratories or new students have a tendency to "reinvent the wheel" for a while, before establishing their own way of making measurements reliably. Standardized test procedures, such as those published by the ASTM, are invaluable, of course, but do not cover all types of tribological measurements. For this reason, we have decided to start a new section of *Tribology Letters* entitled "Tribology Methods", in which we will publish peer-reviewed papers that describe the best way to make particular tribological measurements, as well as introducing new techniques that might be generally useful to the tribological community.

Arguably one of the most misused weapons in the tribological arsenal is the atomic force microscope (AFM). Scott Perry, of the University of Florida, has, with his colleagues, devoted one of our first Tribology Methods papers to this topic [1]. The AFM, which was developed in the 1980s, brought nanometer-scale imaging and later nanoNewton-scale friction measurement within the reach of virtually all tribology laboratories. AFM has the potential to model the friction of a *single*

22

asperity with a counter-surface, to provide information on *real* contact areas, and to provide mechanistic insights into friction on the molecular level. Unfortunately, the literature abounds with contradictory tribological data obtained by AFM, and reliable comparisons between

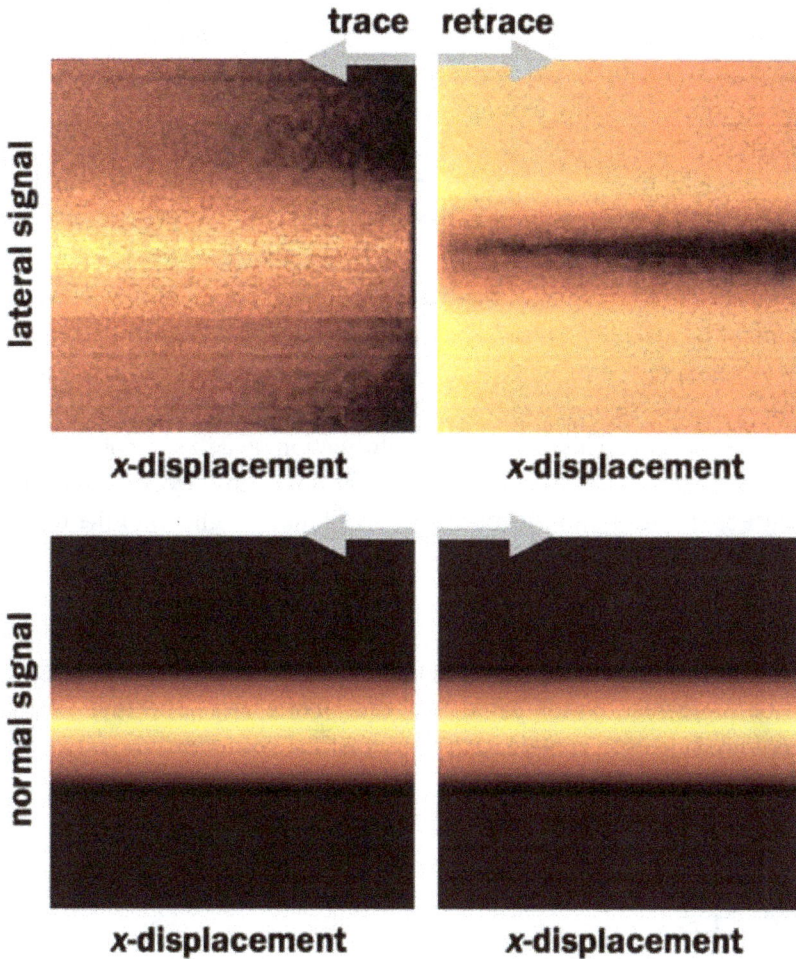

Fig. 1 Normal and lateral signal images generated from AFM by rastering the tip across the sample from left to right (trace) and vice versa (retrace), as load is applied and removed. The lateral images show contrast between the traces as the friction force vector is always opposite to that of the direction of motion. By kind permission of Springer Science+Business Media from Reference [1].

laboratories have been difficult, at best. As a result, AFM friction data have yet to be routinely applied for design purposes. One particular problem has been the calibration of the lateral force constant of the AFM spring cantilever (the force-measuring "heart" of the instrument), although a number of techniques have been described in the literature for doing this, and are comprehensively referenced by Perry. A further, less-well-documented issue, is the misalignment of the laser beam (used for detecting the cantilever motion), which also leads to uncertainties in the measured friction values; Perry and his colleagues have deliberately misaligned a laser beam in their AFM to show how this can have dramatic effects on friction measurements. One of the most useful sections of the paper is the detailed description of how to carry out friction-load measurements with an AFM. Instead of recording friction at a number of discrete load values, an approach that is frequently reported in the literature, Perry and coworkers suggest that *continuously* varying the load while scanning along a line on a sample generates a useful *friction-load map*. This can provide information on adhesion phenomena, hysteresis effects between loading and unloading cycles, and non-linearities in the friction-load curve — yielding insights into the contact mechanics. Additionally, some information can be gained into the onset of wear.

We hope that our new Tribology Methods section will contribute to a higher quality of data in the literature, as well as leading to results that can be more readily compared between laboratories. The first issue of *Tribology Letters* to contain such papers will appear shortly.

Tribology and Lubrication Technology
April 2009, 65(4) p56

Further Reading:

[1] Limpoco, F.T., Payne, J.M., and Perry, S.S. (2009). Experimental Considerations when Characterizing Materials Friction with Atomic Force Microscopy, Tribology Letters, 35, pp. 3–7.

Experiment and Theory, Rubbing Along Together

A special issue of Tribology Letters highlights collaborations between experimentalists and theorists.

In the first of these columns, we pointed out that Newton's laws of motion and Amontons' laws of friction were published at almost the same time (at the end of the 17th century) and that, while Newton's initial laws have evolved into other theories such as relativity and quantum mechanics, the understanding of Amontons' laws, by contrast, has been much slower. This is, in part, because of the difficulties in interrogating a sliding, solid-solid interface and because of the complexity of the processes occurring at that interface. However, the intervening decade since the publication of that first article has seen remarkable advances in both our experimental and theoretical capabilities.

Tribological experiments are now routinely carried out with exquisite delicacy by means of atomic force or friction force microscopes and also in well-controlled, ultrahigh vacuum environments, to obtain well-characterized surfaces that are free of contaminants. Tribometers have been designed not only to measure friction forces but also to monitor the nature of the interface either *in situ* or just after rubbing.

At the same time, theoretical methods, such as molecular-dynamics simulations and first-principles quantum calculations now allow the complexities of the processes occurring at the sliding interface to be modeled with unprecedented precision. This growth has in part been due to the ready availability of vast computing power and the development of parallel codes that allow calculations to be carried out simultaneously on a large number of processors.

Experimentalists and theorists at the 1927 Solvay Conference.

We have arguably now reached the point at which theorists and modelers can analyze the systems for which detailed experimental data can be obtained. Combined with these advances has been a growth in the range of analytical tools that can explore the nature of the surface after sliding. These range from focused-ion-beam (FIB) techniques that can be used to obtain information on how the sub-surface region has evolved, to surface-analytical methods such as X-ray photoelectron spectroscopy that can provide detailed information on the surface composition.

With increased experimental and theoretical sophistication inevitably comes increased specialization. However, scientific advances arise from a close interplay between experiment and theory. In the spirit of encouraging closer collaborations between theory and experiment, Tribology Letters is publishing a special issue entitled "Combining Experiment and Theory in Tribology" which is being guest co-edited by Ashlie Martini from the School of Engineering at the University of California-Merced.

The issue includes papers on a wide array of topics, ranging from contact mechanics of skin and textured surfaces to friction of elastomers and step-edges on graphite. It addresses a number of theoretical approaches, from quantum theory of tribochemistry and sliding friction to molecular-dynamics simulations of third-body formation. Other topics include surface roughness, the properties of films formed from zinc dialkyldithiophosphate, and elastohydrodynamic properties of fluid interfaces.

We hope that this special issue will spur further collaborative efforts between experimentalists and theorists. As a final note, all of the papers are available free of charge to STLE members on their web site (at http://www.stle.org/research/tribology_letters.aspx?) by logging into the member's account.

Tribology and Lubrication Technology
June 2013, 69(6) p88

Further Reading:

Special Issue: Combining Experiment and Theory in Tribology, (2013). Tribology Letters; 50(1)

Remembering David Tabor

David Tabor was one of the founders of scientific tribology, and his insights and philosophy remain an invaluable contribution to our field

David Tabor, who died on November 2005 at the age of 92, was one of a handful of people who can be credited with the founding of tribology as a true science [1]. In fact, not only the subject, but also the word can be attributed to Tabor, who coined the term "tribophysics" in the late 1940s, to describe the activities of his research group.

Tabor's impact on our field is comparable to Shakespeare's on the English language. Exploration of the contact between surfaces was the topic of Tabor's PhD Dissertation (1936), and the basis for his first joint paper with his PhD advisor, F.P. Bowden, in 1939. Together, they developed the concept of a small, real contact area, and likened contact between solids to "turning Austria over and placing it on top of Switzerland". Tabor and Bowden were to collaborate for over thirty years, until Bowden's death in 1970, producing innumerable papers that totally changed the way in which people thought about many other fundamental tribological concepts, such as friction of metals and non-metals, frictional heating, boundary lubrication, adhesion, sliding wear, and hardness. Bowden and Tabor published two books that summarized many of the studies to come out of their laboratory in Cambridge: *Friction and Lubrication of Solids*, Part 1 (1950) and Part 2 (1964) [2], which are classic works, and despite their age, essential reading to all embarking on research in tribology. Another of Tabor's influential monographs *The Hardness of Metals,* has recently been reprinted and remains an important text in the field.

David Tabor
1913–2005

Tabor was also active in many other areas, including polymers and colloids. The surface forces apparatus (SFA), which has led to many observations of profound significance for our understanding of tribology fundamentals, as well as phenomena in colloid science, adhesion, biology and geology, was initially developed in his "Physics and Chemistry of Solids" (PCS) group at the Cavendish Laboratory in Cambridge. PCS spawned many researchers who are now world-renowned scientists themselves. Stories abound of Tabor's patient and friendly discussions with PhD students, postdocs and visitors, even following his retirement from the Cavendish in 1981.

David Tabor, already in his early eighties, was one of the founding Editorial Board members of *Tribology Letters,* and wrote the foreword, entitled *Bridging the Gap,* for the first issue in March, 1995 [3]. He described the unfortunate gap between the engineering aspects of tribology, and the more scientific aspects that were emerging from physics and chemistry. He felt that discoveries in surface science could benefit tribology engineers, but there was a problem in that the two communities did not read each other's literature. He hoped that the journal would be a forum for communication that would redress the balance. His credo, already written some thirty years earlier, was that "the contribution the physicist can make to tribology will be greatly increased by effective and enlightened collaboration with the chemist, metallurgist and engineer". This interdisciplinarity, which could by no means be taken for granted at that time, was one of Tabor's great strengths, and has greatly benefited the field of Tribology since his time. David Tabor was exceedingly helpful in advising us during the startup phase of our new journal. Friendly, wise, gentle, and totally without a shred of arrogance, he assisted us in our choice of Editorial Board members, and in setting the course for Tribology Letters, which has essentially remained unchanged over the last decade.

Tribology and Lubrication Technology
August 2006, 62(8) p64

Further Reading:

[1] Briscoe, B. and Hutchings, I. (2006). David Tabor 1913–2005, Tribology International, 39, pp 591–592.

[2] Tabor, D. and Bowden, F.P. (2001). The Friction and Lubrication of Solids. Oxford University Press.

[3] Tabor, D. (1995). Bridging the Gap, Tribology Letters, 1, pp. iv–v.

David Tabor's photograph is courtesy of Prof. John Field, Cavendish Laboratory, Cambridge University.

Tribute to the Late, Great Mike Gardos

We pay tribute to a great colleague and friend, Michael Gardos

Cutting Edge articles are intended to bring new fundamental developments in tribology to the attention of TLT readers. However, this month, we would like to change the emphasis and highlight the work of an outstanding tribological scientist, Dr. Mike Gardos. While Mike made many outstanding contributions to tribology in the areas of lubricious oxides and lubrication in space environments, to name just two examples, one of his major achievements was in furthering our understanding of the frictional properties of silicon.

Silicon is used for the fabrication of microelectromechanical systems (MEMS), which aim to integrate extremely small (in the micrometer range) mechanical elements, sensors and actuators on a silicon chip. Such MEMS devices have a wide range of potential applications and are currently being used, for example, as accelerometers for crash air-bag deployment systems in automobiles. This has allowed discrete accelerometers to be replaced by MEMS devices, which are integrated with their associated electronics on a single silicon chip, thereby significantly reducing the cost from $50 to about $5 per automobile. Silicon is used for MEMS applications since the technology developed for fabricating integrated electronic circuits on silicon is extremely well developed.

Unfortunately, the lifetime of silicon in MEMS devices is short because of its excessive wear, an effect that is exacerbated by its high

31

coefficient of friction. Mike was the first to realize the origin of this effect. The silicon atoms in the solid, unlike those in metals, are bonded covalently. That is, the bonding electrons in silicon are relatively localized between the atoms rather than being free to move throughout the solid, as in a metal. This, of course, endows silicon with its semiconducting properties.

Mike Gardos, 1938–2003.

Unfortunately, when silicon is cleaved, these bonds are broken and form so-called "dangling bonds" at the surface. Mike pointed out that, when two silicon surfaces are brought together, the dangling bonds interact, resulting in the high friction coefficients and wear rates found experimentally. Under certain conditions, the atoms at the silicon surface can adjust their positions — a process known as reconstruction — so that the dangling bonds now interact with each other. Mike realized that such a reduction in the number of dangling bonds should result in a reduction in friction and wear and showed experimentally that this was the case.

Another way to prevent the dangling bonds from interacting with each other is by reacting them with, for example, hydrogen, and Mike demonstrated that hydrogenating silicon also substantially reduced friction and wear, further confirming his ideas. Mike also pointed the way to a solution to this problem and suggested that diamond would provide a much more suitable MEMS material, since it is readily passivated to yield a low friction coefficient and has a wear rate 10,000 times lower than silicon. Mike was also a regular contributor to Tribology Letters, and one of its first and most enthusiastic editorial board members.

Sadly, Mike passed away just over a year ago, and his insights and creativity will be sadly missed. As a tribute to Mike's career and contributions to tribology a special symposium being organized by Dr. Said Jahanmir will be held at the STLE Annual Meeting in Toronto.

Tribology and Lubrication Technology
June 2004, 6(4) p64

Further Reading:

[1] Gardos, M.N. (1996). Surface Chemistry controlled Tribological Behavior of Silicon and Diamond, Tribology Letters 2, pp. 173–187.

[2] Gardos, M.N. (1996). Tribological Behavior of Polycrystalline and Single-crystal Silicon. Tribology Letters 2, pp. 355–373.

[3] Spencer, N.D and Tysoe. W.T. (2003). Obituary: Mike Gardos. Tribology Letters 15, pp. 1.

[4] Fleischauer, P. (2003). A Tribute to Dr. Michael N. Gardos. Tribology Letters 17, pp. 349–350.

2013 Sees the Passing of Two Eminent Tribologists

We remember the careers, the contributions, and the warm personalities of two tribology pioneers, Sanjay Biswas and Brian Briscoe

We mourn the loss of two colleagues, Sanjay Biswas and Brian Briscoe. Both were ardent supporters of Tribology Letters and served on our Editorial Board, Brian being the very first to agree to sign on to our fledgling journal, back in 1995.

Sanjay Biswas was born in Calcutta, India, graduating in mechanical engineering from the Indian Institute of Technology in Kharagpur. He obtained his Master's degree from the University of Strathclyde, Scotland, in 1969, going on to get his PhD from the University of Birmingham, England, in 1972. In 1976 he joined the Mechanical Engineering Department of the Indian Institute of Science (IISc), in Bangalore, where he served in many capacities, including Dean of Engineering.

Sanjay's research ran the gamut from surface forces and fundamental studies of PTFE wear, to very practical solutions for the oil and automobile industries. He fearlessly entered many areas of tribology, surface science and even cell biology during his career, often generating strikingly original results, but always with an impeccably strong theoretical underpinning. His scientific legacy is a deeper understanding of many aspects of wear, lubrication, and materials behavior, as well as having established a generation of Indian tribologists.

Sanjay was a true visionary, not only in science, but also in his social activism and his work for the IISc, where he established a new undergraduate program, a new focus for the institution on bioengineering, and tirelessly championed those not typically represented in the Indian higher-education scene. Highly respected in the Indian Government, the Institute, and in the tribology community, among his friends he was enormously entertaining, with a wicked sense of humor.

Sanjay Biswas
1945–2013

Brian Briscoe was born in Yorkshire, England and obtained his degrees at the University of Hull, and the University of Cambridge in the Cavendish Laboratory. In 1970, he took a position as Assistant Director of Research/Oppenheimer Fellow at Cambridge, working extensively with David Tabor, and in 1978 became a Lecturer in Interface Science. In 1984, he moved to the Chemical Engineering Department at Imperial College as a Reader in Interface Science, and was promoted to Professor in 1992. He retired in 2009, but stayed at Imperial College as an Emeritus Professor until his passing in June this year.

Brian carried out seminal work on boundary lubrication and the origins of friction in the late 1970s, research that has recently become of renewed importance with the need for low-friction lubricants. He then spent much of his career carrying out pioneering work on the mechanical, friction and wear properties of polymers and this culminated in the publication of the book, "Polymer Tribology" with Sujeet Sinha in 2009. However, Brian's research interests were wide-ranging, including ceramics, pastes, polymer adsorption, condensation, composite materials and ionic liquids; research that is described in his more than 200 published papers. Brian was the founding editor of the journal Tribology International, where he served for 17 years. While Brian will be remembered for his seminal scientific achievements, his students will remember him as a caring supervisor, who worked and enjoyed life with them as he would his own friends. His Christmas parties were the most eagerly awaited events of the year.

Brian Briscoe
1945–2013

Both of these larger-than-life individuals — their friendship, their wise counsel and their invaluable insights — will be missed by us, and by the many people whose lives they touched.

Tribology and Lubrication Technology
February 2014, 70(2) p64

Topic 2

Fundamentals of Friction and Damage

This Chapter on *Fundamentals of Friction and Damage* is by far the largest, with fifteen articles. This partly reflects the emphasis of the *Cutting Edge* column, but also our own biases. The articles can be broadly classified into fundamental experimental methods and theoretical strategies for understanding friction and wear.

The best known fundamental nanoscale experimental technique is the atomic force microscope (AFM), based on the invention of the scanning tunneling microscope — the first scanning-probe method — by Binnig and Rohrer in 1981. The AFM was rapidly extended to being able to measure both normal and lateral forces exerted between a small tip and a surface. Very early work from Tabor's group had shown that real surfaces came into contact at the tips of many small asperities. The ability to understand the contact and frictional properties of a single tip held the promise of being able to simplify the problem of surfaces that have a multiplicity of contacts.

The AFM has been used to examine how frictional energy is dissipated into the vibrations of substrate atoms. This was achieved by changing the vibrational frequency of hydrogen atoms by isotopically substituting them for deuterium and finding changes in friction (*Good vibrations?* February 2008). In *Shake, rattle and slide*, October 2006, AFM was also used to show how rapidly vibrating the tip while sliding, significantly reduced friction and stick-slip motion, suggesting that this trick could be used in microelectromechanical devices.

While the AFM does simplify the contact, as the *Contact Conundrum* chapter emphasizes, precisely determining the contact area, even in such simple systems, is still a challenge. A strategy for addressing this was discussed in *Nudging nanoscale objects*, October 2010 where a nanoparticle on the surface is pushed along the surface, rather than sliding a tip.

A variant of AFM, nanoindentation, which uses larger forces to measure local hardness, has been used to measure the variation in the mechanical properties of soot particles (*The diesel dilemma*, December 2011) that inevitably make their way into the lubricating oil of diesel engines.

The AFM has also been used to explore one of the most challenging problems in tribology of measuring the temperature at a buried interfacial contact in *Moving heat in nanocontacts*, August 2014. This work identified a curious phenomenon of quantized thermal conductivity at the atomistic contacts between a flattened, silicon AFM tip and an amorphous carbon substrate, by measuring the temperature change as the tip came into and out of contact with the substrate.

A completely different approach that examines the collective behavior of the multiplicity of asperity contacts during sliding was described in *Seeing the start of shear*, June 2010. Here, the asperity contacts between two relatively rough transparent blocks of poly (methyl methacrylate) (PMMA) were measured by illuminating the interface and identifying the contacting regions; light was transmitted when the surfaces contacted, but reflected in regions where they did not, providing unique insights into the nature of the contact (as in *The Contact Conundrum Cracked*, February 2007, Topic 7).

This section on Friction Fundamentals also contains intriguing or unusual phenomena that caught our attention. These were covered in April 2011 in an article on the Bauschinger Effect (first described in the mid 19th century), entitled *Changing direction reduces wear*. This effect is manifested as a decrease in yield strength after sliding when the strain direction is changed, and can have important implications for the way in which machines are designed and manufactured; designing machines that operate by maximizing bidirectional sliding should reduce wear, while the abrasive processes used in manufacturing could be improved by unidirectional sliding.

In *Slip-sliding away...*, December 2003, we discussed a quirk in the solution of the Reynolds equation for modeling hydrodynamic lubrication, which we are taught to solve using boundary conditions that assume the liquids do not slide at the interface — the "no-slip boundary condition" (NSBC). It turns out that this is not always valid and that slip can occur when the surface is very smooth and not wetted by the liquid.

The transition from hydrodynamic to mixed and boundary lubrication that occurs when the separation between the surfaces decreases is conventionally described using the Stribeck curve. Lubrication in the boundary regime relies on chemically modifying the surface using additives, but the shape of the Stribeck curve in this region has never been very well understood. *Left of the Stribeck Curve*, in December 2012, described how carefully modifying the structure of the additive could be used to tune the shape of the Stribeck curve in this region.

Another dogma was dispelled in polymer science, as discussed in *Dogma run over by karma*, April 2005. The conventional wisdom in polymer tribology holds that increasing the molecular weight of polymers reduces their propensity to wear. In fact, it turns out that, since wear seems to occur by polymer chains being pulled out of the bulk, this depends on the extent to which they are entangled.

Because of the experimental challenge involved in directly seeing what is happening at a sliding interface, our understanding of what happens at this interface must rely heavily on modeling. As the physicist Paul Dirac pointed out in 1929 "The fundamental laws necessary for the mathematical treatment of a large part of physics and the whole of chemistry are thus completely known, and the difficulty lies only in the fact that application of these laws leads to equations that are too complex to be solved" — a quote that can easily be applied to tribology. The laws that he was referring to are those of quantum mechanics. While this was true with the (non-existent) computing power available in 1929, this is much less true today. In *Theoretical friction modeling*, June 2008 we described how quantum theory can be used to describe the friction of oxides and sulfides of molybdenum and the sliding between ice and boron nitride.

However, quantum mechanics can still only be applied to relatively simple systems, and capturing the complexities of real sliding interfaces relies on combining the potentials derived from quantum mechanics with classical equations of motion using molecular-dynamics (MD) simulations, as we discussed in *Modeling molecular motion*, February 2004.

MD simulations have been used to gain mechanistic insights into a range of tribological problems. In *Why does Amontons' law work so well?*, August 2004, we discuss how MD simulations for a roughened gold surface lubricated by a liquid hexadecane film reveal that there is a broad local distribution of loads at the nanoscale, with an extremely non-linear dependence of friction force on load; Amontons' law is not obeyed locally. However, they appear to average out to give a linear relationship between friction force and normal load for all but the lowest loads.

MD simulations have also been used to explore structural changes at a sliding interface, and *Analyzing high-speed sliding*, December 2008 reported on the way that the interfacial structure of metal-metal interfaces evolves during sliding. Perhaps the most worrying aspect for experimentalists is that the simulations revealed that after the sliding motion has ceased, the surface structure changes significantly from that present while the surfaces were still sliding.

One view of frictional processes sees friction in atomic-scale sliding as arising from the energy needed to surmount the potential energy barrier while the atoms at the interface slide over each other. *Dissipative dislocations*, June 2007, explored how the energy dissipation during sliding can instead be treated as being due to the motion of dislocations and can be used to provide simple, analytical equations for the friction force based on the well-known behavior of dislocations in the bulk.

Good Vibrations?

Replacing the hydrogen on silicon or diamond surfaces by heavier deuterium allows the effect of vibrational frequency on friction dissipation to be measured

As Joule demonstrated in the 19th century, frictional work converts translational kinetic energy into vibrational energy in the form of heat. A key issue, however, is how the energy is *initially* dissipated during sliding.

This could occur either by directly exciting the lattice vibrations or by electronic excitations that are ultimately converted into vibrational energy (heat). Energy dissipation *via* vibrational excitation can be thought of as occurring in the same way that energy is transferred between any two colliding objects, where the energy transferred per collision depends on the ratio of the masses of the colliding objects.

In the case of friction, this is the ratio of the mass of the contact to the mass of the atoms at the surface. The overall rate at which this occurs also depends on the number of times the atoms in the contact collide per second and therefore on the vibrational frequency of the atoms at the surface.

In principle, this idea can be tested experimentally by changing the vibrational frequency of the surface and exploring whether this has an effect on friction. The problem, of course, is that changing the atoms at the surface will also change the chemical composition of the sample.

Professor Rob Carpick, now at the University of Pennsylvania, has come up with an elegant solution to this problem by isotopically modifying surfaces of diamond or silicon by terminating them with either hydrogen or deuterium [1]. Deuterium is twice as heavy as hydrogen, but is identical in all other aspects (Fig. 1). This allowed them to make chemically identical carbon and silicon samples that were terminated with hydrogen or deuterium and to compare their frictional properties by means of atomic force microscopy.

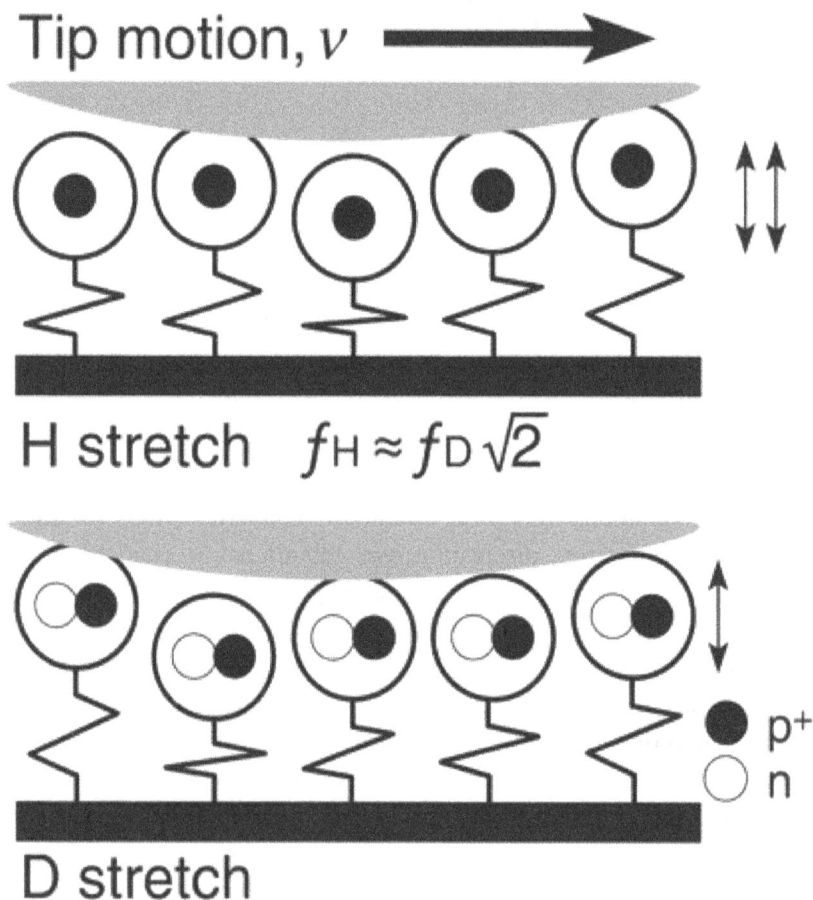

Fig. 1 A schematic of the frictional interface. Vibrating adsorbates collide with and dissipate kinetic energy from the moving tip at a rate that depends on the adsorbate's frequency and thus its mass; that is, at different rates for H than for D. From Reference [1]. Reprinted with kind permission from AAAS.

A key to being able to perform this experiment successfully was confirming that they indeed had deuterium or hydrogen at the surface, and that these exhibited different vibrational frequencies. In order to achieve this, Dr. Steve Baldelli at the University of Houston measured the sum-frequency generation spectra of the samples.

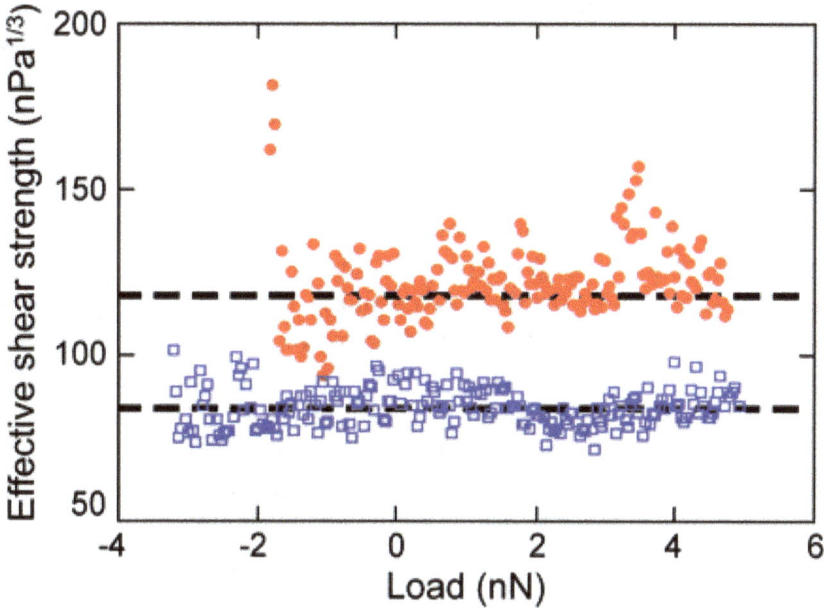

Figure 2: Average effective shear strength (dashed lines) on Si(111) and residuals from representative fits for the Si-H (solid red symbols) and Si-D (open blue symbols) measurements. From Reference [1]. Reprinted with kind permission from AAAS.

This experiment involves simultaneously shining laser beams of visible and infrared radiation onto the samples. These frequencies can combine to emit radiation at a frequency that is the sum of the incident frequencies — hence the name sum-frequency generation. This is a very low-probability process and therefore requires high-intensity lasers.

This mechanism is enhanced when the surface vibrates at the same frequency as the incident infrared laser, so that measuring the sum-frequency intensity as a function of the frequency of the tunable infrared laser yields a vibrational spectrum. The key to this experiment is that, because of the symmetry properties of the process, it is only sensitive to surface vibrations and not to those in the bulk. The results confirmed that the silicon and carbon samples could be exclusively covered by either hydrogen or deuterium and that the hydrogen-terminated surface vibrated at a frequency of 8.5×10^{13} per second and the deuterium-covered surface at a lower value of 6.5×10^{13} per second as expected.

Carpick's group showed that the work of adhesion of the deuterium- and hydrogen-terminated surfaces was the same, implying that they were chemically identical. However, the friction, measured by atomic force microscopy, was different for the different samples (Fig. 2).

Since the friction force also depends on the contact area, this was estimated by calculating the contact area for an adhesive contact, which allowed an interfacial shear strength to be extracted. To avoid possible problems of different tip geometries, friction was measured by repeatedly going back and forth between two samples using the same tip. This showed that the shear strength of hydrogen-terminated carbon was about 1.26 times higher than that of the deuterium-terminated surface, demonstrating that energy is dissipated directly by exciting vibrations at the surface. Similar results were found for silicon, where the ratio was about 1.3. These experiments elegantly demonstrate that surface vibrations are important in the friction of silicon and diamond. Whether this applies for all interfaces still remains to be fully established.

Tribology and Lubrication Technology
February 2008, 64(2) p72

Further Reading:

[1] Cannara, R.J., Brukman, M.J., Cimatu, K., Sumant, A.V., Baldelli, S., Carpick, R.W. (2007). Nanoscale Friction Varied by Isotopic Shifting of Surface Vibrational Frequencies, Science, 318, pp. 780–783.

Shake, Rattle and Slide

Modulating an atomic force microscope tip causes the friction force to decrease to zero

The microelectronics industry has developed a myriad of strategies for shaping and forming silicon into complex electronic devices. Over the last few years, efforts have been made to use the methods developed by this industry to make small mechanical devices, or micro electromechanical systems (MEMS), from silicon.

However, while silicon has exactly the right properties for making integrated circuits, MEMS devices are prone to serious adhesion and wear problems. Traditional liquid lubricants cannot be used in the very small spaces present between the MEMS components. Other approaches, such as chemically modifying the silicon surface or gas-phase lubrication have met with limited success.

An alternative strategy has been suggested by Ernst Meyer's group at the Department of Physics and Astronomy of the University of Basel, Switzerland, working in collaboration with Roland Bennewitz at the Department of Physics, McGill University, Canada [1]. Rather than trying to develop methods of lubricating the sliding surfaces, they have tried to physically perturb the system such that friction itself disappears (Fig. 1). While this may seem like an unlikely strategy, they recently demonstrated that it might in fact be feasible.

In their experiment, they slid a sharp tip over the basal plane of either a sodium chloride or potassium bromide single crystal, which had been cleaned in an ultrahigh vacuum. The experiment was carried out using an atomic force microscope, where the sharp tip was attached to a silicon cantilever, and the normal and lateral forces were measured from the deflection or torsional movement of the cantilever. These were determined very precisely by reflecting a laser beam from the back of the cantilever onto a position-sensitive light detector.

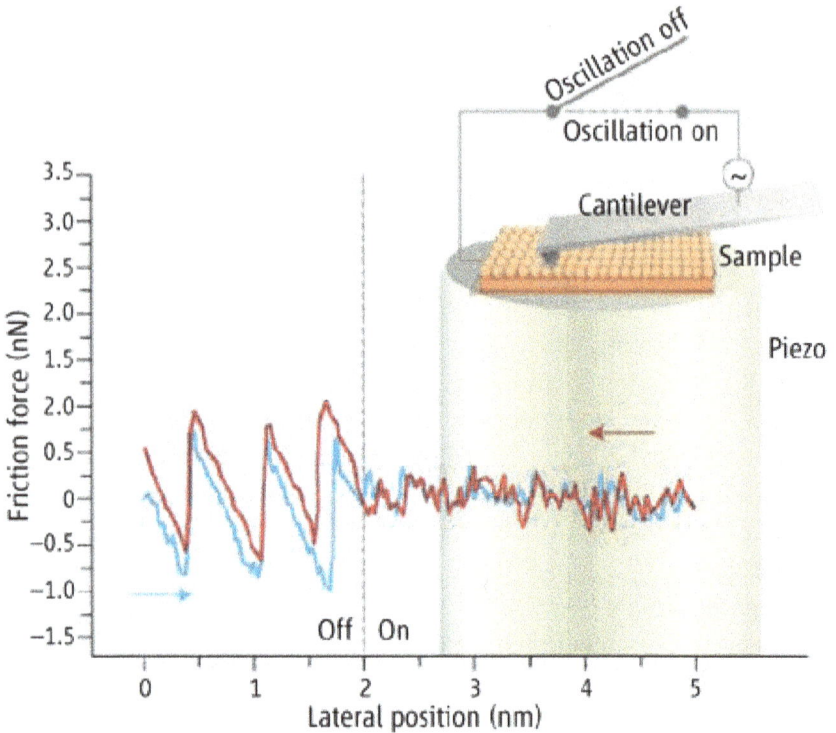

Fig. 1 Friction in a nanometer-scale contact, in the form of atomic-scale stick-slip instabilities (left), is dramatically reduced (right) when a modulation in the normal force is applied to the interface (sketch, top right). Reproduced by kind permission from Reference [2].

As expected, frictional forces were measured, and the apparatus was sufficiently sensitive to detect stick-slip motion with a periodicity equal to the atomic spacing on the surface. Meyer and co-workers then applied an a.c. voltage between the sample holder and the silicon tip. This causes the tip to vibrate and they found that the frictional force was reduced when the oscillation frequency was in resonance with the vertical vibrations of the contacting tip, but not with the lateral vibrations.

They further found that the friction decreased as the amplitude of the applied a.c. voltage increased until it reached a point at which friction disappeared altogether (at about 1 V), after which, of course, no further decrease in friction was found (Fig. 2).

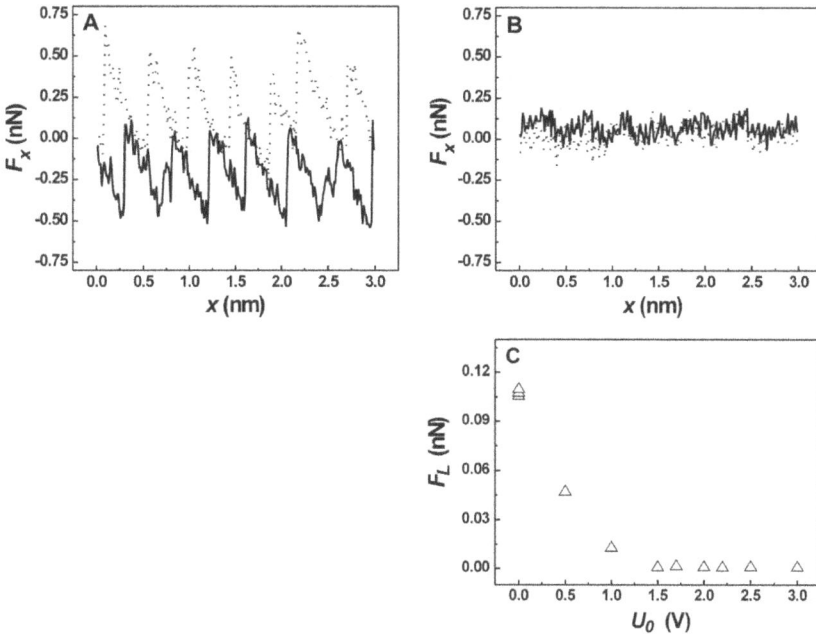

Fig. 2 Lateral force detected by scanning forward and backward on an atomically flat NaCl surface. An average normal load F_N = 2.73 nN was kept constant by a feedback loop (A) without a bias voltage between the cantilever and the sample holder and (B) with an applied ac voltage with frequency f = 56.7 kHz and amplitude U_0 = 1.5 V. In (A) and (B), the solid curves refer to the forward scan, whereas the dotted curves refer to the backward scan. (C) Average frictional force (the mean value of F_x in either scan direction) as a function of the voltage amplitude U_0. From Reference [1]. Reprinted with kind permission from AAAS.

The authors provided an atomic-level explanation for the phenomenon. The stick-slip motion that they measured without oscillating the tip arises from the variation in potential energy as atoms in the sharp tip slide over atoms at the sample surface. As the cantilever is initially deflected laterally, the atoms at the surface are trapped at the minimum of this potential. As the cantilever continues to move, it forces the atoms to move up this potential energy curve, until a point is reached where the cantilever has moved a sufficient distance that the atoms slide to the maximum of the potential energy curve, and they then slide rapidly down to the next minimum in the potential, leading to the observed atomic stick-slip motion. The role of the normal vibration is to facilitate this atomic sliding so that it occurs at very much lower lateral forces.

The authors suggest that the parts in physical contact in MEMS devices are likely to resemble the contacts in their atomic force microscopy experiments and it is relatively easy to apply a few volts to such devices. Unfortunately, it is unlikely that this method can be generally applied to macroscopic systems because there will be a wide range of resonances in realistic contacts.

Tribology and Lubrication Technology
October 2006, 62(10) p80

Further Reading:

[1] Socoluic, A, Gnecco, E., Maier, S., Pfeiffer, O., Baratoff, A., Bennewitz, R., and Meyer, E. (2006). Atomic Scale Control of Friction by Actuation of Nanometer-Sized Contacts, Science 313, pp. 207–210.

[2] Carpick, R.W. (2006). Controlling Friction, Science, 313, pp. 184–185.

Nudging Nanoscale Objects

A 'Tip-on-top' method allows for more precise positioning
and measurement of interfacial friction

The conventional approach to measuring nanoscale friction using an atomic force microscope (AFM) is to bring a tip down onto a surface and to measure the lateral force during sliding, from the torsion of the cantilever. While this method is capable of precisely measuring miniscule friction forces, it can suffer from disadvantages due to uncertainties in the tip structure or cleanliness.

Over the past few years, an alternative approach has been used, in which a nanoparticle deposited onto the surface is moved by an AFM tip Fig. 1). This allows the friction of a wide range of materials to be measured on clean, well-defined surfaces. This approach also offers the possibility of fashioning nanostructured arrays (Fig. 2). It would seem that the obvious way to carry out this experiment is to place the AFM tip next to the nanoparticle and simply push it to a desired location. It turns out that, unless the tip is placed exactly at the center of the nanoparticle, the particle has a tendency to move sideways, making it difficult to precisely manipulate it.

This issue has been addressed with experiments performed by Dirk Dietzel in the group of André Schirmeisen at the University of Munster in cooperation with Udo Schwarz at Yale University [1]. They developed a simple geometrical model for pushing nanoparticles that calculates the particle trajectory as a function of the relative impact parameter (the distance of the tip from the center of the particle, ratioed to the particle diameter). The simple model is verified experimentally using antimony as well as gold nanoparticles, and they find, for example, that for a relative impact parameter of only 5%, the particle has been completely pushed aside after moving only three times the particle radius.

To solve this problem, they suggest an alternative strategy; "tip-on-top" manipulation. The authors theorize that, if the frictional force between a tip placed on top of a nanoparticle is larger than the force required to slide the particle on the substrate, it will simply move with the tip. At the same time, the cantilever torsion can be interpreted as a measure of the interfacial friction between the nanoparticle and the substrate. They tested this idea using antimony nanoparticles on a highly oriented, pyrolytic graphite substrate, and found that, indeed, they could precisely manipulate the antimony particle at will, and that it accurately followed the motion of the AFM tip if the tip exerts sufficient load.

'Tip on Side'-Approach

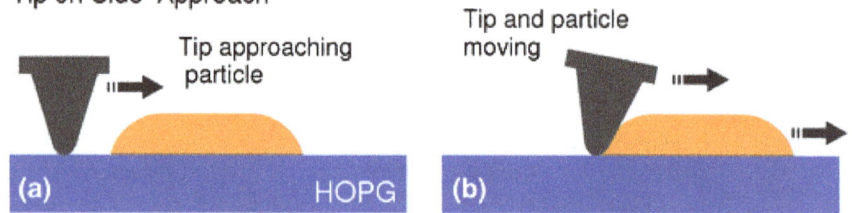

'Tip on Top'-Approach

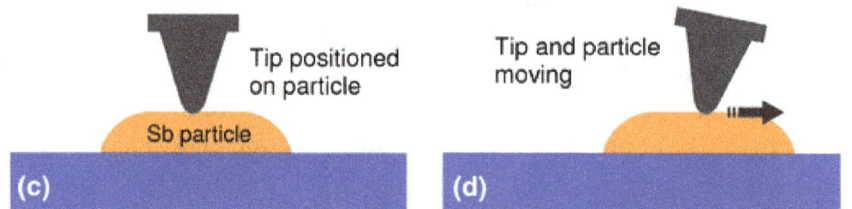

Fig. 1 Schematic representations of the two manipulation approaches. 'Tip on side' (top row): first the tip is positioned besides the nanoparticle and approaching it (a), once the tip hits the particle it starts pushing the nanoparticle along its path (b). 'Tip on top' (bottom row): The tip is first positioned on top of the nanoparticle (c), if the interfacial friction between tip and nanoparticle is sufficiently high, the tip can be moved and the nanoparticle will follow (d). By kind permission of Springer Science+Business Media from Reference [1].

This raises the question as to whether the additional load exerted by the AFM tip will influence the interfacial friction. To test whether this was the case, the lateral force required to push the particle in the conventional way was measured. In this case, the lateral force to slide the tip on the graphite surface served as a reference, and the increase in force, of ~62 nanoNewtons (nN), when the tip encountered the sample, was taken to be the interfacial friction.

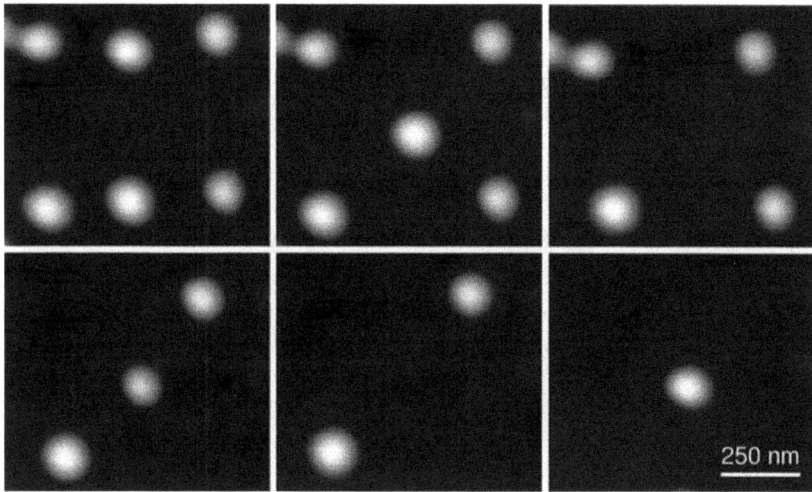

Fig. 2 Demonstration of precise nanoparticle positioning of a group of Sb nanoparticles, which have been arranged to resemble the pips of a dice. First, a group of nanoparticles has been placed to form the number 6. After that, the nanoparticles have been removed from the scan area one at a time and the remaining nanoparticles were rearranged to resemble the numbers 5–1. Manipulations have been done *via* vector paths using the 'tip on side' approach. By kind permission of Springer Science+Business Media from Reference [1].

In the case of the "tip-on-top" method, there is no such internal reference. In order to solve this problem, the authors measured the difference in the forces needed to push the particle forwards and backwards. This is taken to represent twice the friction force, leading to a value of 64nN — in excellent agreement with that measured by pushing. Indeed, no influence of normal force was found in any of the friction measurements. This observation was rationalized by arguing that the cantilever normal forces (of ~10 nN) resulted in only very small increases in pressure between the nanoparticle and substrate (<1 pN/nm^2) so that the interface is not changed.

This strategy not only provides a precise method for nanoparticle manipulation, but it also allows friction to be measured on the same particle as a function of a wide range of parameters, such as sliding velocity, and sliding direction. It could even be applied to investigate the onset of nanoparticle wear processes.

Tribology and Lubrication Technology
October 2010, 66(10) p88

Further Reading:

[1] Dietzel, D., Feldmann, M., Herding, C., Schwarz, U.D. and Schirmeisen, A. (2010). Quantifying Pathways and Friction of Nanoparticles During Controlled Manipulation by Contact-Mode Atomic Force Microscopy, Tribology Letters, 39, pp. 273–281.

The Diesel Dilemma

Studying soot on the nanometer scale helps us understand its role in engine wear

Reducing harmful emissions from automobile engines always involves tradeoffs. In Diesel engines, one of these is the choice between soot production and the formation of nitrogen oxides (NOx). In the diffusion flame that characterizes the Diesel combustion process, increasing the amount of oxygen leads to soot reduction and greater NOx emission, while reducing oxygen consumption reduces NOx emission, but increases soot. This is sometimes known as the "Diesel dilemma". Since NOx and soot emissions from automobiles are both highly regulated, it is almost inevitable that there will be some degree of soot produced by modern Diesel engines. Of course soot does not only end up in the tailpipe, but a significant fraction enters the lubricating oil, and that is where our tribological story begins…

There is a considerable literature on the effects of lubricant soot content on engine wear, and the field is complex. On the one hand, there may be simple abrasion of some engine components by hard soot particles, while on the other hand, soot has been reported to interact chemically with antiwear additives, either by adsorbing them, reducing their availability to the wearing surfaces, or by removing them from the contact regions. Simultaneously, the soot can even have a beneficial effect on wear, increasing oil viscosity and therefore film thickness. The latest efforts to understand the behavior of soot [1] have been reported in STLE-affiliated journal *Tribology Letters* by Hiralal Bhowmick and

Prof. Sanjay Biswas of the Indian Institute of Science in Bangalore, India [1].

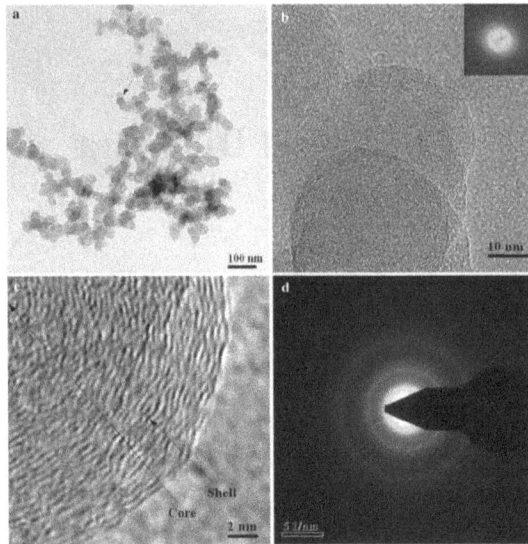

Fig. 1 a) Low-magnification image of a particle agglomerate, b) and c) HRTEM of the particle marked with an arrow in a), d) representative selected area electron diffraction of a soot particle. By kind permission of Springer Science+Business Media from Reference [1].

Bhowmick and Biswas have created a model soot by burning ethylene, but what makes their study unusual is that they have extracted individual soot particles from different parts of the flame and measured their sizes and crystallinity by high-resolution transmission electron microscopy, hardness by nanoindentation, and tribological properties by lateral force microscopy (LFM) in an atomic force microscope (AFM). In order to be able to correlate these properties with behavior in other flames and with other reports in the literature, the researchers also carried out careful temperature measurements at different parts of the flame.

The particles appeared to be around 21 nm in diameter, being slightly larger in the cooler parts of the flame. However, the morphology varied significantly, from completely amorphous at the root of the flame (temperature 1665 K) to amorphous but surrounded by a columnar crystalline shell (which was graphitic, with long-range order) in the hottest part of the flame (1689 K). As the particles move into the cooler part of the flame (1657 K), the amorphous core gradually shrinks, and the majority of the particle consists of small crystallites embedded in an amorphous phase.

Nanoindentation measurements, which were carried out by holding the particles in position on a silicon wafer with a 10 nm PMMA layer, revealed that the hardness of the particles increases significantly (from 3 to 4.5 GPa) upon going from the hottest to the coolest part of the flame, while the friction coefficient (as measured by a DLC-coated AFM tip) appears to increase concomitantly. By repeated scanning of the AFM tip over individual particles, the amount of removed material as a function of applied load could be determined. While this measurement was only semiquantitative, it was found that the particles taken from the hottest part of the flame, i.e. those with the graphitic shell, were most prone to material removal, presumably due to the weak interlayer bonding.

The hardness of the hardest soot particles formed in these model experiments (3-5 GPa) was not so far away from that of cast-iron engine liners (5-7 GPa), meaning that abrasive wear of the liners is a possibility, and wear of the additive-produced layers even more likely. Clearly these careful experiments indicate that soot particles can span a relatively wide range of morphologies — and therefore different mechanical and tribological properties — and that there is not one single answer to the question of how soot affects engine wear. Flame temperature, in particular, seems to play an important role in tribological behavior of the soot. Further experiments, involving soot-additive interactions, are now clearly needed.

Tribology and Lubrication Technology
December 2011, 67(12) 80

Further Reading:

[1] Bhowmick, H. and Biswas, S.K. (2011). Relationship Between Physical Structure and Tribology of Single Soot Particles Generated by Burning Ethylene, Tribology Letters, 44, 139–149.

Seeing the Start of Shear

Measuring the evolution of the real area of contact over short timescales provides insights into the onset of sliding

We previously reported in this column on an elegant method for measuring the real area of a static contact that was developed by Professor Jay Fineberg and his group at the Hebrew University of Jerusalem in Israel [1]. The interface between two contacting Plexiglas blocks was illuminated with a laser beam. The light reflected from regions that were not in contact, but was transmitted at the asperity-asperity contacts, thereby enabling the real area of contact to be measured from the amount of light transmitted.

The Fineberg group has more recently refined the experiment to address the much more complicated issue of what happens to the interface when a shear force is applied [2]. Now, the real area of contact is imaged in real time using a fast video camera. An additional refinement is that displacements of the block are simultaneously measured by focusing another laser beam onto grids attached to the side of the moving block. Motion is detected by monitoring the diffracted beam using a precise position-sensitive detector.

An experiment is carried out using blocks of poly (methyl methacrylate) (PMMA) with a root-mean-square roughness of about 1 micrometer. A normal load is applied to the blocks and a lateral force is then slowly applied using a load cell. The contact is imaged and the block displacement simultaneously monitored. Increasing the lateral, shear force leads to the appearance of a series of detachment fronts that propagate from the trailing edge of the contact (where the force is

56

applied). Macroscopic slip finally occurs when the detachment front has reached the leading edge of the block.

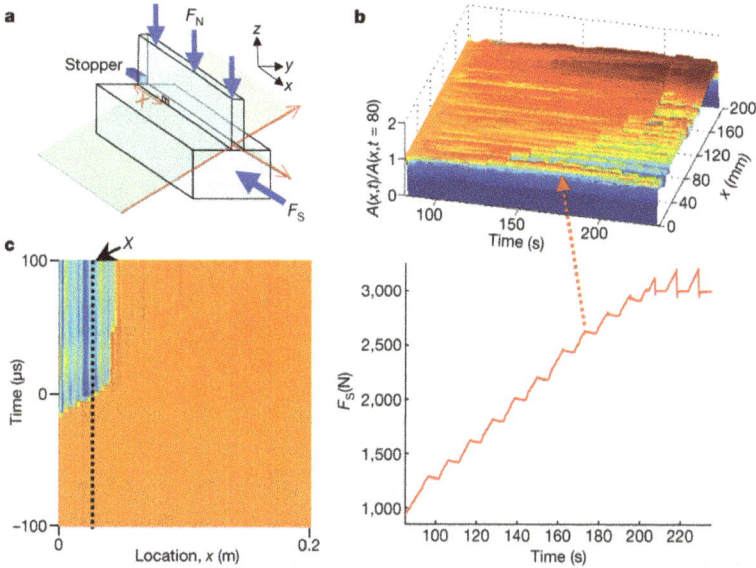

Fig. 1 a, Schematic diagram of the experimental system. A uniform normal load F_N is applied to two PMMA blocks (base and top blocks). Shear force F_S is applied in the x direction to the base block, which is mounted on a low-friction stage. F_S is countered by a rigid stopper at the slider's x = 0 edge. The real area of contact $A(x, t)$ throughout the contact plane is imaged at 4-μs intervals. Simultaneous measurements of slip, both at a distance X from the stopper and at the top block's leading edge were made. b, The non-spatially uniform shear loading generates a sequence of rapid precursors before the onset of overall motion, which initiate at x = 0 and arrest at successively larger distances within the interface. The top panel shows $A(x, t)$ normalized by $A(x, t = 80$ s$)$ taken before the first precursor. Hotter colors (reds) denote increased $A(x, t)$. Colder colors (blues) denote reduced $A(x, t)$. Each precursor significantly reduces $A(x, t)$ along its length. The bottom panel shows the corresponding loading curve $F_S(t)$. On acoustic detection of each precursor, the continuous increase of F_S in time is paused for 5–1,000-s intervals. Slip, generated by each precursor at x = 0, gives rise to small sharp drops in F_S that are detectable due to the compliance of the loading system. Here, $F_N = 6,000$ N and $\mu_S = 0.51$, and the origin of t was the time at which F_S was applied. c, Short-time measurements of $A(x, t)$ (at 4-μs intervals) reveal that each precursor is a detachment front propagating along the interface at velocities v approaching the Rayleigh-wave speed $c_R = 1,280$ m s^{-1}. Here v = 1,200 ± 100 m s^{-1}. The dotted line denotes the location X where slip $\delta(t)$ was measured. Here, $A(x, t)$ was normalized by its value at 1 ms before the front's passage. Reprinted by kind permission from Macmillan Publishers Ltd: Nature (Reference [2]), copyright (2010).

Insights into the sequence of events that occur during slip are obtained by simultaneously measuring the time dependence of the real area of contact and the displacement. Four distinct regimes are identified. In the first regime, lasting a few microseconds, there is an approximately 20% reduction in the real area of contact, but no net slip takes place. It is suggested that this is when interfacial fracture takes place. The energy thus dissipated causes a significant temperature rise that weakens the PMMA-PMMA interface, which has a glass-transition temperature of about 110°C.

A second phase is observed, which lasts about 60 microseconds, where the contact area remains constant, but rapid slip takes place at sliding velocities between 5 and 20 cm/s. This is proposed to be due to a weakening of the strength of the contact due to the fracture-induced temperature rise and the characteristic time for this phase is just that required for the interface to cool to ambient temperature.

A third, slower phase is then seen, where the slip velocities lie between 0.1 and 2 cm/s, which lasts for a few hundred microseconds. The real area of contact still remains constant and this region is ascribed to interfacial slip, which is slowed because of the presence of stronger contacts at lower temperatures. Slip ceases completely after this, and the contacts then reform.

While the authors acknowledge that the interpretation of the various regimes found for PMMA sliding against PMMA is strictly only relevant to glassy materials, they conjecture that a similar, fracture-induced weakening mechanism might also apply to other systems such a brittle materials like granite where, in this case, rather than thermal softening of PMMA, weakening might occur by the crushing of interlocking asperities.

Tribology and Lubrication Technology
June 2010, 6(6) p88

Further Reading:

[1] Tysoe, W., and Spencer, N. (2007). The Contact Conundrum Cracked, Tribology and Lubrication Technology, 63(2), pp. 64.

[2] Ben-David, O., Rubinstein, S.M., and Fineberg, J. (2010). Slip-stick and the Evolution of Friction Strength, Nature, 463, pp. 76–79.

Changing Direction Reduces Wear

The extent of abrasive wear between a pair of metal counterfaces seems to be greater for unidirectional sliding than when sliding reciprocally. The reason seems to lie in the Bauschinger Effect

It has long been known that the stress-strain properties of materials can depend on the way in which stresses are distributed microscopically within the material. This effect is known as the Bauschinger Effect, named after Johann Bauschinger (1834–1893), who was a mathematician and Professor of Engineering Mechanics at the Technical University of Munich.

The Bauschinger Effect is frequently observed as a decrease in yield strength of metals when the strain direction is changed, and can be explained in terms of dislocation buildup in the material, which leads to strain hardening, and is reduced when the strain direction is reversed due to local reversible movement of dislocations and annihilation of dislocations with opposite signs.

The Bauschinger Effect has consequences for tribology, as has recently been reported by C. Y. Tang, D. Y. Li, and G. W. Wen of the University of Harbin in China and the University of Alberta in Canada [1]. These researchers carried out wear experiments in a modified ASTM G65 tester by rotating a steel wheel loaded to 130 N against a copper-zinc alloy surface. During a given set of experiments, which were carried out for a fixed number of rotations, they varied the number of times the direction of rotation of the wheel was changed and looked for the influence of this quantity on wear, hardness, and local structure, as measured by X-ray diffraction (XRD) and scanning electron microscopy (SEM).

Fig. 1 Johann Bauschinger, 1834–1893.

The results showed that unidirectional sliding (i.e. no direction reversal of the wheel) led to up to 40% higher wear than cases where the rotation direction was reversed. Beyond five cycles of changing direction, the effect did not increase significantly. Overall, the Vickers' hardness decreased as the number of direction-change cycles was increased. XRD measurements of unidirectionally *versus* bidirectionally

(50 cycles) worn surfaces showed negligible differences, suggesting that textural changes in the alloy did not play a major role in the wear differences. On the other hand, subsurface cross-sections measured by SEM showed that the unidirectionally worn samples had suffered more severe fracture than the bidirectionally worn materials. Fracture involves the nucleation and propagation of cracks, which occur when stress concentrations exceed the critical stress at fracture. The stress concentrations arise when dislocations of the same sign interact with each other, piling up at interfacial boundaries. Due to the Bauschinger Effect, local reversible movement of dislocations may occur and the number of dislocations is reduced in bidirectional sliding, as dislocations become annihilated by dislocations of opposite sign, which arise from sliding in the opposite direction. Furthermore, bidirectional sliding may not only reduce crack formation, but it may also slow crack propagation, since the stress driving the crack is also reversed.

There are interesting consequences of these observations. Firstly, design of machines that minimize unidirectional in favor of bidirectional sliding could lead to reduced wear. Secondly, abrasive processes used for manufacturing could be improved if unidirectional sliding were favored over bidirectional sliding.

Tribology and Lubrication Technology
April 2011, 67(4) p56

Further Reading:

[1] Tang, C.Y., Li, D.Y., and Wen, G.W. (2011). Bauschinger's Effect in Wear of Materials, Tribology Letters, 41, pp. 569–572.

[2] Sowerby, R., Uko, D.K (1979). A review of certain aspects of the Bauschinger effect in metals, Mater. Sci. Eng., 41, pp. 43–58.

[3] Yue, L., Zhang, H., and Li, D.Y. (2010). Defect generation in nano-twinned, nano-grained and single crystal Cu systems caused by wear: A molecular dynamics study, Scripta Materialia, 63, pp. 1116–1119.

Slip-Sliding Away

The no-slip boundary condition is fundamental in bearing design, but does it always correspond to reality?

Why don't we get clean by simply standing under the shower? The velocity of the flowing water may be a meter per second or so away from the body, but near the skin surface it's much lower, and therefore needs help from a hand or washcloth (and soap, but that's another story) to get us clean. Similar phenomena thwart us when we try to blow dust off a mirror or rely on rain to clean our cars.

The idea that flowing liquids actually come to a standstill at a solid interface, otherwise known as the "no-slip boundary condition" (NSBC) is also inherent in engineering tribology: The NSBC is assumed in the derivation of the Reynolds equation, and is therefore fundamental to many of the bearing-design equations that we learn. These equations, of course, work pretty well. Nevertheless, recent investigations using techniques such as the surface forces apparatus (SFA) by Steve Granick at the University of Illinois [1], and Olga Vinogradova at the Max Planck Institute for Polymer Research in Mainz, Germany [2] suggest that even with simple, one-component liquids, "wall-slip" can occur if the surface is very smooth and not wetted by the liquid. This deviation from the NSBC is quantified as the "slip-length", which is defined as the notional distance *inside* the solid surface at which the liquid appears to come to rest, if the velocity profile is extrapolated across the interface. Slip lengths greater than one micrometer have been observed in the SFA.

Why does slip happen? One possibility is that liquids inherently form a low-viscosity layer at interfaces that they do not wet, which then leads to an apparent, non-zero slip length. Another possibility is that a thin layer of gas or vapor is formed at the non-wettable interface, and it is within this layer that the slip occurs. Which model is correct under which conditions remains a highly active area of research.

Hugh Spikes at Imperial College, London, has recently demonstrated the implications of wall-slip on bearing design [3,4]. In normal bearings, the wetting of the surfaces by the lubricant facilitates its entrainment into the contact region. Without wetting, there would be a possibility of starvation. On the other hand, the wetting phenomenon, with the consequent upholding of the NSBC, leads to a viscous ("Couette") drag on the moving surfaces, which is chiefly responsible for the friction of the bearing. In the case of a pair of non-wetted surfaces, and thus the presence of wall-slip, this drag would be reduced, but on the other hand the lubricant would no longer be able to support the load on the bearing. The ideal case, according to Spikes, would be to construct a "half-wetted" bearing, with one non-wetted and one wetted surface, the latter moving with respect to the convergent zone. In this case, under thin-film, low-load conditions (which are of tremendous significance in micromachines and in disk-drive systems) the friction of the bearing would be considerably reduced over the traditional, wetted, no-slip case.

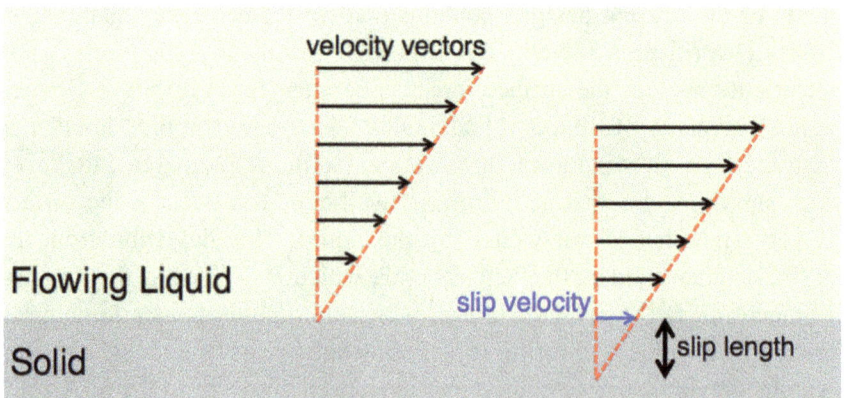

This story shows that fundamental tribological research, such as that carried out with the SFA, can lead us to insights yielding new developments in engineering tribology. It also shows that such insights come to tribologists who have a familiarity with both worlds.

Tribology and Lubrication Technology
December 2003, 59(12) p64

Further Reading:

[1] Granick, S., Zhu, Y. and Lee, H. (2003). Slippery questions about complex fluids flowing past solids, Nature Materials, 2, pp. 221–227.

[2] Vinogradova, O.I. (1999). Slippage of water over hydrophobic surfaces, Int. J. Miner. Process., 56, pp. 31–60.

[3] Spikes, H.A. (2003). Slip at the wall — evidence and tribological implications, *Proc. 29th Leeds/Lyon Symposium, Tribological Research and Design for Engineering Systems,* ed. D Dowson *et al.,* Elsevier, Amsterdam pp. 525–535.

[4] Spikes, H.A. and Granick, S. (2003). Equation for slip of simple liquids at smooth solid surfaces, Langmuir, 19, pp. 5065–5071.

Left of the Stribeck Curve

Adsorbed molecules influence the way lubricated contacts change their friction with speed

We are all familiar with the Stribeck curve, which describes the way in which friction varies with viscosity, speed, and load (captured by the Sommerfeld number, $\eta U/W$) in a lubricated contact. At high values of the Sommerfeld number, the behavior is well understood, and describes well-known, rheological effects, but the behavior at low values generally remains shrouded in mystery. Many textbooks simply show this boundary-lubrication region as a horizontal straight line, with the friction locally unaffected by the value of the speed, load, or viscosity. Reality differs from this picture.

In the boundary region, where the contacts are close and chemistry rules supreme, effective lubrication generally relies on the use of additives. These compounds adsorb on the sliding surfaces and provide a low-shear-strength interface, where sliding takes place with less friction and wear than it would be on the hard, asperity-ridden surfaces immediately beneath.

It turns out that the molecular structure of the additives plays a significant role in determining the shape of this boundary region of the Stribeck curve. In a recent paper [1] in the STLE Journal *Tribology Letters*, Sophie Campen and Hugh Spikes, of Imperial College London, collaborating with Jonathan Green, Gordon Lamb, and David Atkinson of Castrol, Ltd., have revisited the issue, both reviewing existing reports, and generating interesting, new data. Many previous studies had shown that friction increases with sliding speed in lubricants containing organic friction modifiers, such as stearic acid. Such behavior has important consequences, since a positive slope in this region of the Stribeck curve guarantees that stick-slip will not occur (thus avoiding clutch shudder in automatic transmissions, for example). The new experiments explored the boundary region, using a variety of additives dissolved in hexadecane. While all chains were 18 carbons in length, some contained double bonds. While straight-chained, saturated stearic acid showed the expected positive slope in the boundary region of the curve, the unsaturated oleic acid showed a slight negative slope (see illustration above). The issue of whether it was the double bond itself or the bent shape of oleic acid that led to this difference was resolved by examining elaidic acid. This molecule, like oleic acid, contains a single double bond, but in the so-called *trans* configuration, which leads to a straight-chained molecule. In the Stribeck curve, elaidic acid behaved like stearic acid, showing that the straightness of the chain, rather than the double bond itself, and presumably the ability of the molecules to pack into ordered monolayers, were influencing the sign of the slope.

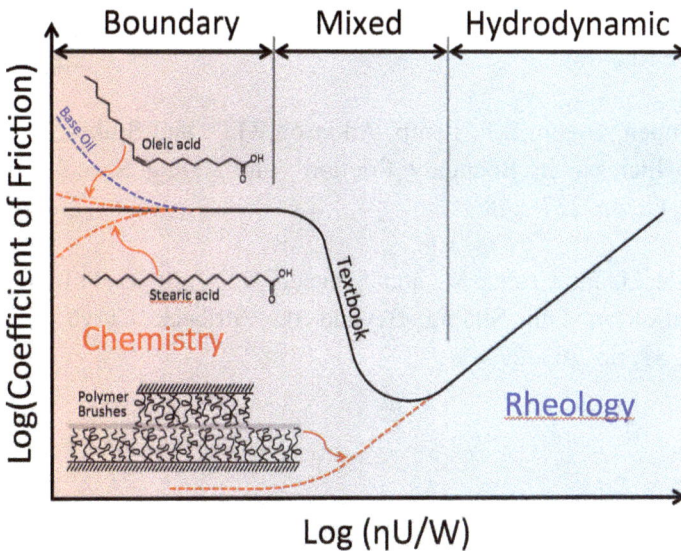

Fig. 1 The Stribeck Curve, as shown in textbooks, and as shown recently to be the case for oleic acid, stearic acid, and polymer brushes.

Another approach to modifying the low-Sommerfeld-number region of the curve was recently reported [2] in *Tribology Letters* by Robert Bielecki, Maura Crobu, and Nicholas Spencer, at the ETH Zurich, Switzerland. In this work, closely-spaced, oil-compatible polymer chains (collectively referred to as a "polymer brush"), several hundred nm in length, were grown out of silicon and steel surfaces. The behavior under a variety of oils differed markedly from that of the bare surfaces (and the textbook Stribeck curve), in that the hydrodynamic region seemed to extend much further to the left and provided lower friction coefficients, finally leveling off at a roughly constant value. In contrast to hard-hard interactions, even when modified with small-molecule additives, it appears that soft contact between oil-swollen brushes provides an intrinsically low value of friction, presumably provided by a readily sheared, oil-rich layer at the outer edges of the contacting polymer brushes.

Tribology and Lubrication Technology
December 2012, 68(12) p96

Further Reading:

[1] Campen, Green, S.J., Lamb, Atkinson, G.D. and Spikes, H. (2012). On the Increase in Boundary Friction with Sliding Speed, Tribology Letters, 48, pp. 237–248.

[2] Bielecki, R., Crobu, M. and Spencer, N.D. (2013). Polymer-Brush Lubrication in Oil: Sliding Beyond the Stribeck Curve, Tribology Letters, 49, pp. 263–272.

Dogma Run Over by Karma

Considerations of fundamental aspects of polymer science lead to new insights into why some polymers have lower wear than others

There's a dogma in polymer tribology that goes "to reduce wear you have to increase molecular weight". Based on this, in low-wear applications, such as liners for pipelines, the choice is generally ultrahigh molecular weight polyethylene (UHMWPE). Unfortunately UHMWPE has an exceedingly high melt viscosity and therefore cannot be processed by common polymer processing methods, such as injection molding; instead, one must begin by sintering UHMWPE powder and then machining the fused mass as if it were a block of wood. This is not only costly and cumbersome, but also means that the finished product can show signs of incomplete particle fusion as well as machining marks.

Over the years, a number of parameters have been suggested to correlate with the abrasive wear resistance of polyethylene. While some researchers have concluded that it is the number-average molecular weight, $\bar{M}_n$ that determines abrasive wear properties, others have found the weight-average molecular weight, $\bar{M}_w$, to be the critical parameter. Still others have tried to correlate abrasive wear with melt viscosity or crystallinity of the polymer. Theo Tervoort, Jeroen Visjager, and Paul Smith (TV&S), working at the Department of Materials of ETH Zurich in Switzerland, have carried out abrasive-wear measurements on a large library of different linear polyethylenes, spanning three orders of magnitude in molecular weight and a wide range of polydispersities [1]. They also obtained accurate values for the molecular weight distribution and crystallinity, and plotted the wear values against these parameters. No satisfactory correlations were found with any of these properties.

At this point, TV&S went back to the "molecular drawing board", reasoning that the abrasive-wear process involved chains being pulled out from the bulk, rather than being broken, which would be a much more energetically demanding process. The difficulty in pulling out the chains increases with their length, but it also specifically depends on the degree of "entanglement" of the chains with each other. This concept has been worked on extensively in the past, and for each chain there is thought to be a mean distance between entanglements ("molecular knots") with other chains where the "entanglement distance" can be expressed as a molecular weight, found to be about 1250 g/mol for polyethylene. There must be at least three such entanglements along a chain's length to lead to a load-carrying network, and thus an improvement in wear. This implies that chains must have a molecular weight at least four times greater than the entanglement distance; smaller chains do not contribute to forming a network, but merely act as a diluent. This would seem to imply that wear resistance should correlate with the average number of entanglements per chain, which turns out to scale with the volume fraction of non-diluent chains, ϕ, multiplied by the number-average molecular weight adjusted for the fraction of diluent, $\bar{M}_n^*$. An excellent correlation was found when the reciprocal wear coefficient was plotted against $1/\phi\bar{M}_n^*$, (Figure 1), meaning that the theory had predictive powers.

Interestingly, this correlation implies that high wear resistance should be obtainable not only for ultrahigh molecular weights, but also at moderate molecular weights, providing that the polydispersity is low. Such polymers can be synthesized using modern metallocene-based-catalysts. Since $\bar{M}_w$ dictates viscosity, and therefore melt-processability, one can envision polymers that combine high wear resistance with melt processability, a concept that flies in the face of the old dogma. TV&S have indeed been able to demonstrate exactly this combination of properties for some of the samples they tested.

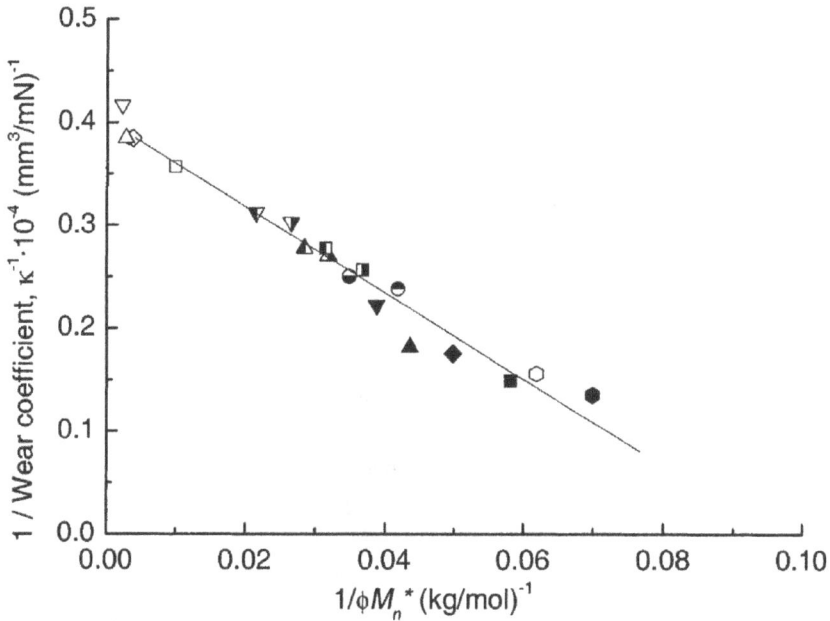

Fig. 1 Graph of the reciprocal wear coefficient as a function of $1/\phi\bar{M}_n^*$ for various polyethylene samples. The line is a least-squares fit of all data. Reprinted with kind permission from Macromolecules. Copyright 2002 American Chemical Society. Reference [1].

TV&S' work shows that gaining a molecular understanding of the problem can inspire new approaches that help transcend old tribological dogmas.

Tribology and Lubrication Technology
April 2005, 61(4) p64

Further Reading:

[1] Tervoort, T.A., Visjager, J. and Smith, P. (2002). On the Abrasive Wear of Polyethylene, Macromolecules, 35, pp. 8467–8471.

Theoretical Friction Modeling

First-principles quantum calculations are used to calculate friction coefficients and model tribochemical reactions

One approach to modeling atomic friction theoretically, as we have discussed in TLT [1], is by classical molecular dynamics simulations of the sliding interface. Here, the potentials used as inputs for the simulations are calculated using quantum mechanics.

Until recently, this approach has been taken since it was prohibitively difficult to calculate realistic potentials for two sliding surfaces because of the large number of atoms involved. The main problem is that individual wavefunctions — the solutions of the Schrödinger equation — have to be calculated for each electron in the system. For large systems, with even larger numbers of electrons, the problem was essentially impossible to solve.

Meanwhile, two important changes have taken place over the last decade or so. First is the availability of faster computers. Second is the realization by physicists that it's not really necessary to describe each electron individually, but that the collective distribution of all of the electrons could be used instead so that the total energy of atoms, molecules or solids depends on the electron density distribution.

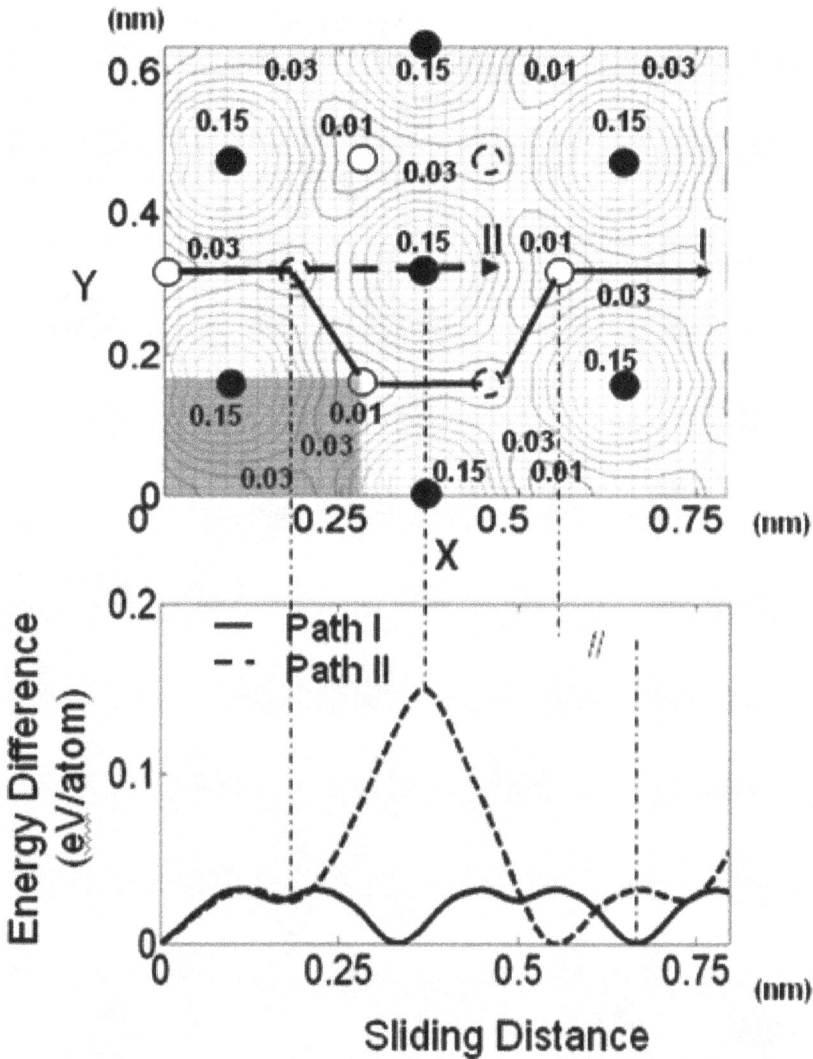

Fig. 1 Static potential energy surface of MoO_3 as a function of position. Top: contour plot of energy difference as a function of displacement in X and Y directions (parallel to the sliding surface). Open circles represent O atoms on the top of sliding interface and solid circles represent O atoms underneath. Bottom: energy difference along the sliding pathways. Reprinted with kind permission from Reference [2]. Copyright (2008) by the American Physical Society.

A function that depends not on a variable but another function — in this case the function that describes the spatial variation of the electron density — is known as a functional, and this method has been dubbed Density Functional Theory (DFT). DFT allows quantum mechanics to be used to describe more complex and realistic systems, in particular the tribological interface.

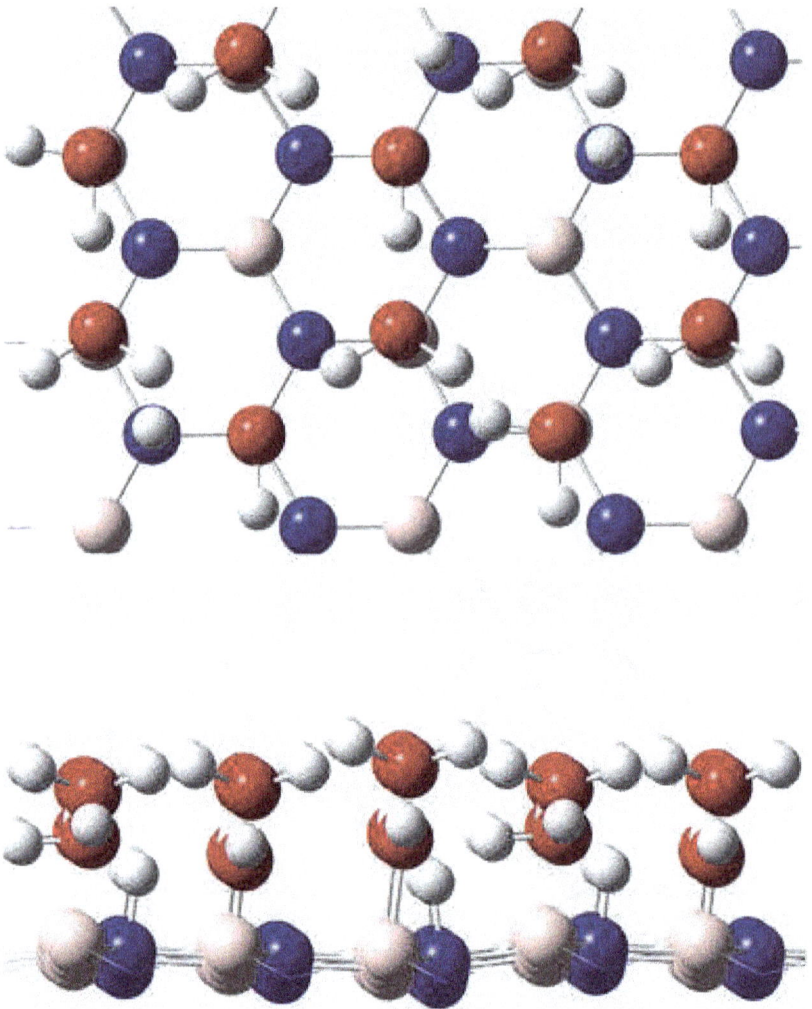

Fig. 2 A tribochemical reaction between the surfaces of ice-Ih and h-BN. Boron atoms are colored pink, nitrogens blue, oxygens red and hydrogens white. By kind permission of Springer Science+Business Media from Reference [3].

Professors Susan Sinnott, Simon Phillpot, Greg Sawyer and Scott Perry of the University of Florida recently used DFT to examine the sliding of two related layered materials, molybdenum disulfide (MoS_2) and molybdenum trioxide (MoO_3), to understand the deterioration in friction of molybdenum disulfide when exposed to air [2].

First, the surfaces were (theoretically) compressed to pressures between 300 and 650 MPa in a direction perpendicular to the layers, and the equilibrium positions of the atoms recalculated yielding strains between 0.1% and 0.4%. DFT was then used to calculate the change in energy caused by moving the surfaces parallel to each other to create a potential-energy landscape

The energetically favored sliding pathway can be determined by looking for the lowest energy trajectory along this landscape. The approach revealed, for example, that the sulfur atoms at the sliding interface in MoS_2 follow a zig-zag pathway. The differences between the energy minima and maxima during sliding are used to calculate the lateral force required to overcome this energy (Fig. 1).

While the authors acknowledge that this is a significant simplification of the real sliding interface, theory predicted that the MoO_3/MoO_3 interface should have the largest sliding friction and the MoS_2/MoO_3 the lowest, in good general agreement with what is found experimentally.

A similar approach has been used by Professor Tapani Pakkanen's group at the University of Joensuu in Finland to examine the friction between ice and hexagonal boron nitride to replicate a water-lubricated interface [3].

He used DFT to calculate a friction coefficient of 0.14, in excellent agreement with the measured values of between 0.06 and 0.14. However, Pakkanen was also able to demonstrate that water bonds strongly to the surface boron atoms, dissociating into NH and B-OH species as the pressure was increased (Fig. 2). Evidently, DFT is set to contribute to our fundamental understanding of friction and tribochemistry, as it has in other areas of physics and chemistry.

Tribology and Lubrication Technology
June 2008, 59(12) p64

Further Reading:

[1] Tysoe, W. and Spencer, N. (2004). Modeling Molecular Motion, Tribology and Lubrication Technology, 60), pp. 80.

[2] Liang, T., Sawyer, W.G., Perry, S.S., Sinnott, S.B. and Phillpot, S.R. (2008). First-Principles Determination of Static Potential Energy| Surfaces for Atomic Friction in MoS_2 and MoO_3, Physical Review B, 77, pp. 104105.

[3] Koskilinna, J.O., Linnolahti, M., and Pakkanen, T.A. (2008). Friction and a Tribochemical Reaction Between Ice and Hexagonal Boron Nitride: A Theoretical Study, Tribology Letters, 29, pp. 163–167.

Modeling Molecular Motion

Molecular-dynamics simulations provide a powerful strategy for modeling sliding interfaces

In principle, all of the theoretical machinery for predicting the chemical, electrical or mechanical properties of materials is available by solving the quantum-mechanical equations of motion, which describe how very small particles — electrons and atoms — move and interact.

Unfortunately these equations are far more difficult to solve than their classical counterparts — Newton's equations of motion. While it is now possible, using modern, high-speed computers, to obtain reasonably accurate solutions for static structures of surfaces, and of adsorbed species on surfaces, the problem of exploring what happens when surfaces move relative to each other is beyond the scope of current theoretical strategies, using even the very fastest computers.

The classical Newtonian equations are much easier (but still not simple) to solve for complex systems. A trick that has been devised to help resolve this problem is to use quantum mechanics to calculate the strength of the forces between atoms and molecules, a problem that can be solved using current theories and computers, and to use these so-called "force constants" to solve Newton's equations of motion to simulate how atoms and molecules move at a tribological interface. This melding of the classical and quantum worlds, dubbed "molecular dynamics", allows theorists to examine, on an atomic scale, what happens when contacting surfaces move relative to each other.

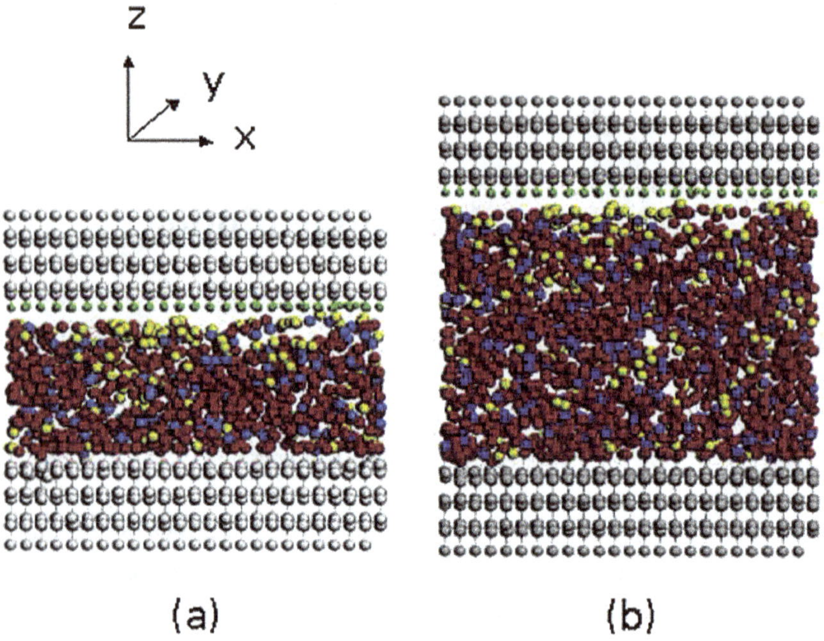

(a)   (b)

Fig. 1 Snapshots of the atomic positions of (a) a thin and (b) and thick films of hydrogen-terminated diamond (111) counterfaces brought into sliding contact with amorphous carbon films attached to the (111) face of diamond. Sliding is along the x direction. Carbon (large spheres) and hydrogen (small spheres) atoms in the diamond substrates are colored gray and green, respectively. Blue, red, and yellow carbon atoms in the films have sp^3, sp^2, and sp hybridization, respectively. Reproduced by kind permission of the American Chemical Society from Reference [2].

Such molecular-mechanics simulations can be used to obtain a general, fundamental understanding of frictional phenomena using simple models for the interactions between atoms. Mark Robbins' group at Johns Hopkins University used such a strategy to demonstrate that the classical Amontons' laws of friction can be predicted by assuming that a layer of adsorbed molecules is trapped between the contacting surfaces [1].

In their simulations, a constant pressure was applied to one solid and a constant lateral force or velocity is imposed along the sliding direction. These simulations found that the local shear stress during sliding varies linearly with the local contact pressure. It was also found that the values of the static and dynamic frictional forces tracked each other, as often found in real-world systems.

Judith Harrison at the United States Naval Academy examined the tribochemistry of alkane molecules trapped between moving diamond surfaces as a function of time using realistic interactions between atoms that can even model the breaking and making of chemical bonds (Fig. 1) [2]. She finds that chemical reactions can occur between the trapped alkanes and the diamond surfaces, including the cleavage of carbon-hydrogen bonds, the recombination of the hydrogen atoms to form hydrogen molecules, and the attachment of the hydrocarbon fragment to the diamond surface. A beautiful movie of this process occurring in real time can be found on the web at:

www.usna.edu/Users/chemistry/jah/MD_movies_images/tribo_sframe.mov

Tribology and Lubrication Technology
February 2004, 60(2) p80

Further Reading:

[1] He, G. and Robbins, M. (2001). Simulations of the Kinetic Friction due to Adsorbed Surface Layers, Tribology Letters, 10, pp 7–14.

[2] Gao, G.T., Mikulski, P.T. and Harrison, J.A. (2002). Molecular-scale Tribology of Amorphous Carbon Coatings: Effects of Films Thickness, Adhesion, and Long-range Interactions, Journal of the American Chemical Society, 124, pp 7202–7209.

Why Does Amontons' Law Work so Well?

Amontons' law * *is empirical, originally observed by Leonardo Da Vinci, and it works for dry sliding surprisingly well. Why?*

A group of scientists led by the theorist Uzi Landman (Georgia Institute of Technology) and the experimentalist Jacob Israelachvili (University of California, Santa Barbara) has published a review of what is known about Amontons' Law in the *Journal of Physical Chemistry* [1]. The authors describe both classical and modern tests of Amontons' Law, and compare experimental values with those obtained from some new molecular dynamics (MD) simulations of both rough and smooth surfaces in contact.

The extraordinary thing about Amontons' Law is that it works on the macroscopic, the microscopic, and nanoscopic scales. Results from conventional tribometers (Fig. 1), the surface forces apparatus (SFA), and the atomic force microscope (AFM) all confirm Amontons' law for non-adhesive contacts. Attempts to explain this ubiquitous behavior have been numerous, but most of us learned about it in terms of the real area of contact (Bowden and Tabor): Friction force (F)$\propto$ Real area of contact

---

*Amontons' law states that for any two materials the friction force is directly proportional to the applied load, with a constant of proportionality, the friction coefficient, that is constant and independent of the apparent contact area, the surface roughness, and the sliding velocity. The law was actually first suggested by Leonardo da Vinci (Figure 2), who lived some 200 years earlier than Amontons.

80

$(A) \propto$ Load (L). But, while $A \propto L$ for plastic deformations, it is not the case for elastic ones, where $A \propto L^{2/3}$ (Hertz). However, for an exponential distribution of asperities, as Greenwood has shown, the relationship conveniently returns to $A \propto L$ for both elastic and plastic contacts. This approach has led to a special place in our thinking about the importance of the real area of contact.

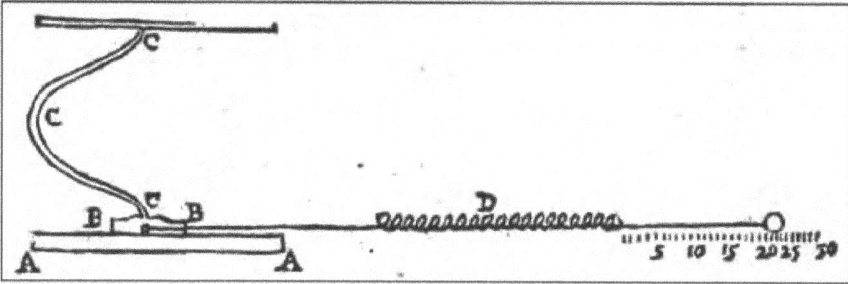

Fig. 1 Amontons' experiments on friction: the spring D was used to measure the friction force during the sliding process between materials A and B. Spring C applied the normal force. From G. Amontons *De frottement de diverses matieres les unes contre les autres (1699)*.

But how special is A, in reality? While the explanation above might hold for a wooden block on a steel plane, it doesn't explain the "single-asperity", non-adhesive AFM or SFA results, where deformation is elastic, but $F \propto L$. The final experimental nail in the coffin comes from simultaneous, direct measurements of F and A as a function of L, between atomically flat mica surfaces in Israelachvili's SFA, carried out under strong brine to eliminate adhesion. The conclusion: friction is proportional to load (i.e. Amontons' Law holds) but *friction is not proportional to the real contact area*!

Further insight comes in the form of molecular dynamics simulations [2] of two sliding, roughened gold surfaces, "lubricated" by a thin, liquid hexadecane film. The resulting F vs L plots for the total system show a linear behavior at all but the very lowest loads, even though the concept of real contact area is never invoked in the simulation, which simply models individual interactions between atoms. However, if the surface is divided into sub-nanometer-scale "tiles", a broad, local load distribution is found (i.e. some atoms are in contact, some are heavily loaded, some not at all), and an extremely non-linear relationship between F and L is observed for individual "tiles" (i.e. Amontons' Law is not *locally* obeyed). However, the data are well fitted by a Weibull Distribution (WD) — a function frequently used to describe failure of materials, where the strength of the weakest component leads to failure.

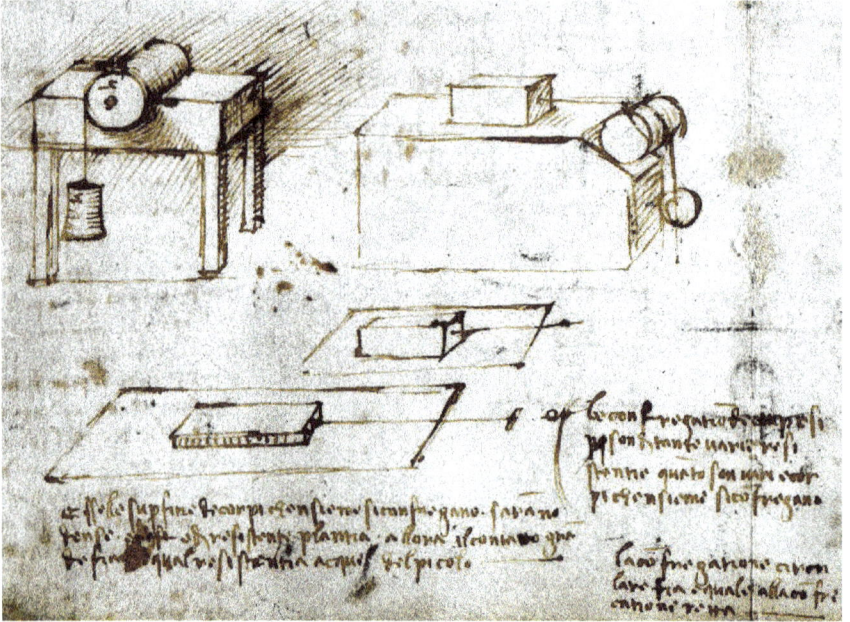

Fig. 2 Illustrations of Leonardo's experiments on friction. From Leonardo da Vinci *Codex Arundel (1508)*.

An analogy is the way in which we can measure the temperature of a gas, while realizing that the individual molecules are all traveling at different speeds (the Boltzmann distribution). In the sliding pair, mechanical energy is being transformed into heat at a constant rate, but locally this rate can vary a great deal. It seems that the Weibull distribution can fulfill a similar role in tribology to that of the Boltzmann distribution in the gas-temperature case: integration over space yields an overall value of friction, which may show considerable local variation. Ultimately the most important factor in determining friction is the number density of atoms, molecules, or bonds that are interacting, and the degree of irreversibility of this interaction. This density and irreversibility fluctuates tremendously in both space and time.

In the end, the article does not explain why Amontons' Law holds. It does, however, show that many assumptions that we make about the way in which friction works are at best oversimplified, or at worst plain wrong, and that a statistical approach is a valuable way to think about the problem.

Tribology and Lubrication Technology
August 2004, 60(8) p64

Further Reading:

[1] Gao, J., Luedtke, W.D., Gourdon, D., Ruths, M., Israelachvili, J.N. (2004). Frictional Forces and Amontons' Law: From the Molecular to the Macroscopic Scale, J. Phys. Chem., 108, pp.3410–3425.

[2] Tysoe, W.T. and Spencer, N.D. (2004). Modeling Molecular Motion, Tribology and Lubrication Technology, 60(2), p80.

Analyzing High-Speed Sliding

Molecular dynamics simulations are increasing our understanding of what happens at the interfacial region

Many tribologists have made the observation that when two materials slide against each other, significant changes occur in the interfacial region. Names for this region have included the *Beilby layer*, the *transfer layer*, the *third body*, and *mechanically mixed material.* Understanding this region, which is known to be highly defected and often contains a significant nanocrystalline component, is crucial, since it is where most of the velocity accommodation takes place, and thus where the temperature is highest and the greatest levels of plastic strain occur. *Post mortem* examination of this layer reveals a composition that is similar to that of the accompanying wear debris.

Recently, Hong Jin Kim and David Rigney, from Ohio State University, working with Woo Kyun Kim and Michael Falk at the University of Michigan, have reexamined this important issue [1]. The researchers utilized molecular-dynamics simulations to model the behavior of two pairs of crystalline materials: self-mated (A-A) and hard-soft (B-A). The model used the well-known Lennard-Jones (L-J) potentials, and defined the hard material as having B-B bonds with an L-J potential well that was twice as deep as those of A-A or A-B bonds. Two blocks of A, or a block of A and B, were brought into contact, slid against each other at a given normal load, and the system monitored as time elapsed. Particular attention was paid to the strain-rate profile, temperature distribution, defect formation, local crystal structure, and the heterogeneity of the material flow.

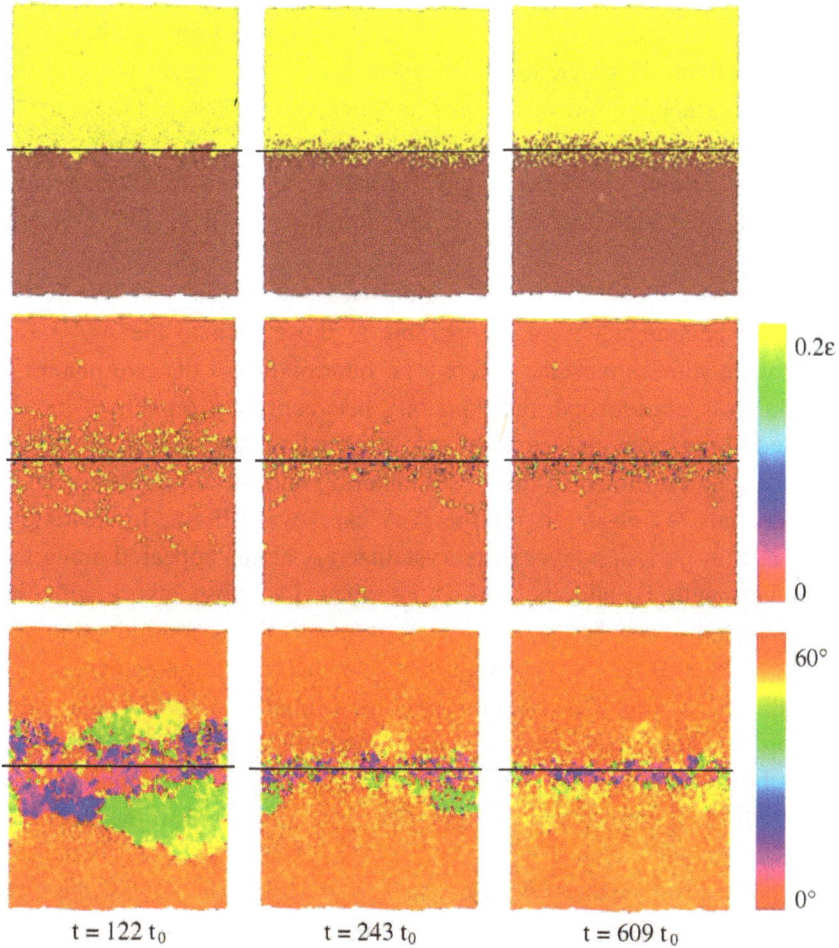

t = 122 t_0 t = 243 t_0 t = 609 t_0

Fig. 1 Snapshots of a sliding self-mated crystal tribopair configuration. The times noted are measured since the beginning of sliding. The line shows the original position of the interface. The top row colors the atoms by whether they were initially in the top or bottom block. The second row shows the defect potential energy. The bottom row shows the local orientation of the crystal structure. By kind permission of Springer Science+Business Media from Reference [1].

In the case of self-mated sliding, initial rapid and dramatic changes occurred, after which the system stabilized. Significant roughening and intermixing of material from the two blocks continued throughout the simulation. Initially, a high defect concentration was found with a broad depth distribution. Later, this was concentrated in a region close to the sliding interface, within which the temperature had exceeded the melting point. Just outside this high-temperature, liquefied region (further away from the sliding interface), the formation of nanocrystallites was observed, their size decreasing as the simulation evolved into a steady state.

The case of hard-soft (A-B) sliding was considered next. While the observed phenomena were comparable to those seen in the A-A case, there were some important differences. Intermixing of the two materials was far less pronounced, the boundary between them remaining highly defined throughout the simulation. Deformation and defect formation were largely restricted to the softer material, and the formation of large nanocrystals, as observed for the A-A case, was absent. Interestingly, after a short period, a layer of recrystallized A atoms appeared along the boundary with B, and the sliding interface, *i.e.* those layers with the highest strain rate that are chiefly responsible for accommodating the velocity, moved to a steady-state position that was actually within the soft sliding partner. This can be viewed as a situation where A is actually sliding on a layer of A, accompanied by some slip at the A/B interface. Such behavior has been observed in dry sliding of aluminum against steel, where the transfer only occurs from the aluminum to the steel and not the other way around.

Another significant result from this work is the behavior of the A-B system upon stopping the sliding and removing the load. As the pressure was removed, and the bulk materials relaxed elastically (*i.e.* expanded), the highly defected area (within A) rapidly crystallized and contracted, accompanied by the elimination of voids. This implies that there is a significant change in the microstructure of the softer partner between steady-state sliding and the *post-mortem* situation — that generally encountered by experimentalists. This could have significant consequences for interpretation of experimental data, and illustrates the need for new approaches that can monitor the sliding interface *in situ*.

Tribology and Lubrication Technology
December 2008, 64(12) p56

Further Reading:

[1] Kim, H.J., Kim, W.K., Falk, M.L., and Rigney, D.J. (2007). MD Simulations of Microstructure Evolution during High-Velocity Sliding between Crystalline Materials, Tribology Letters, 28, pp. 299–306.

Dissipative Dislocations

It is shown that the energy dissipation due to dislocation motion can describe the frictional behavior of metals

Perhaps the simplest, and currently the most commonly used model for explaining friction at the atomic scale was originally proposed nearly eighty years ago by Tomlinson. This describes the shear of an interface as a rigid translation between an atom and a sinusoidal potential that arises as the atom in contact moves from a more stable to a less stable position during sliding.

However, if we envision a contacting interface between two grains of periodic materials, there will generally be interfacial stresses due to mismatch between the two lattices, either because the surfaces are made from different materials or because the lattices are rotated with respect to each other. Such stresses are generally accommodated by generating dislocations.

If the two surfaces are now slid with respect to each other, motion could occur by rigid translations as envisaged by Tomlinson. However, the presence of dislocations provides an alternative mode of interfacial sliding involving motion of dislocations (Fig. 1). Indeed, in bulk materials, plastic deformation almost always occurs by the motion of dislocations rather than the rigid shear of one atomic plane over the other, leading Professor Laurie Marks and Arno Merkle at Northwestern University to suggest that sliding friction should similarly occur via the motion of dislocations [1].

A major advantage of such a description is that the mechanisms of dislocation motion are rather well understood. More importantly, the frictional drag due to dislocation motion can be described analytically using basic materials constants and thus requires no adjustable parameters.

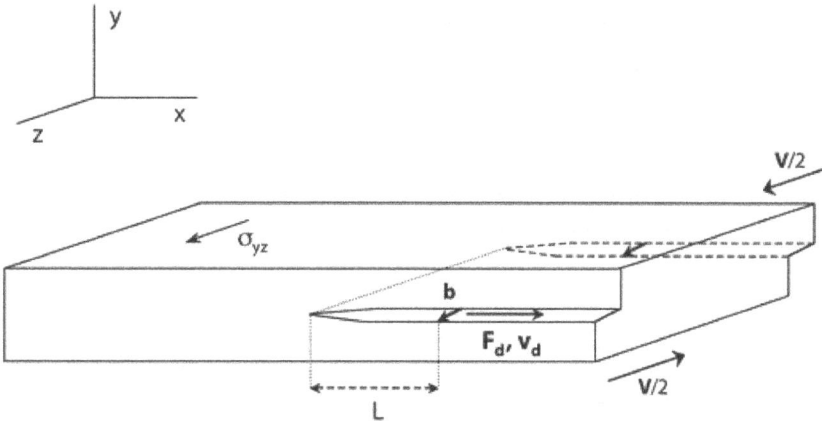

Fig. 1 Force experienced by a screw dislocation resulting from an interfacial shear. By kind permission of Springer Science+Business Media from Reference [1].

Several phenomena have been identified as contributing to dislocation "drag". The first, dubbed "phonon wind" generates lattice vibrations (phonons) through the non-linear elastic properties of the crystal. A second is known as the "flutter" effect, where dislocation motion generates elastic waves and can dominate the "phonon wind" at low temperatures and for harmonic materials. Dislocation motion can also generate electronic excitations, and the net drag is the sum of all of these.

There is a final dissipation mechanism, relevant to low temperatures and the low sliding velocities encountered in most frictional studies, known as "radiation friction". Combining all of these processes leads to a relatively simple analytical expression for the friction force.

The model is compared to experimental results for the sliding of oriented, atomically clean, metal single crystals, where frictional measurements are carried out in ultrahigh vacuum at pressures of around 1×10^{-10} Torr (Fig. 2). This allows perfectly clean metal surfaces to be obtained. The theory predicts that sliding is dominated by radiation friction at low sliding velocities, resulting in a velocity independent friction as found experimentally for sliding velocities below ~0.1 m/s.

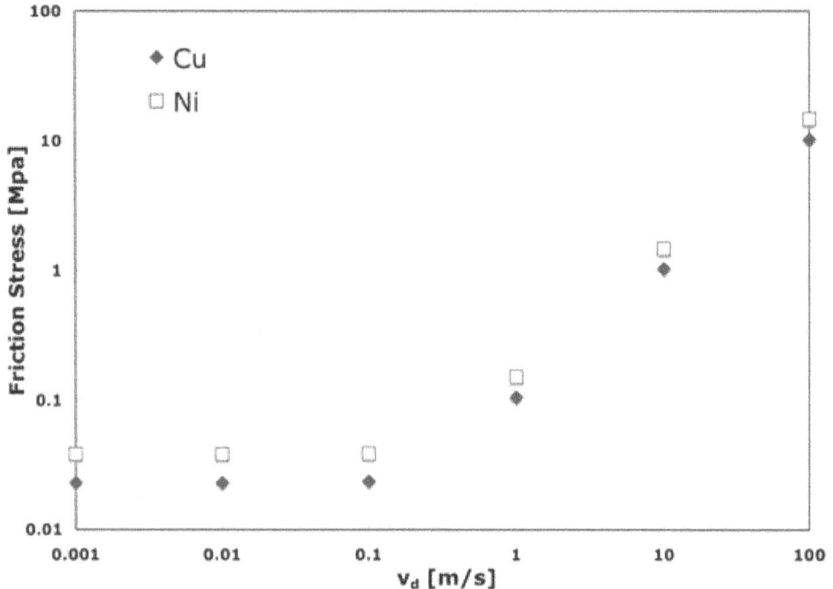

Fig. 2 Frictional stress per dislocation as a function of dislocation velocity calculated at 100 K for Cu(111) and 300 K for Ni(100). The crossover point from pinned to viscous behavior is determined to be 0.1 m/s for the given set of conditions. By kind permission of Springer Science+Business Media from Reference [1].

The friction force is predicted to increase with velocity at higher sliding speeds as other dissipation mechanisms become more important, and this is also found experimentally. Experimental data are available for a number of metal-metal interfaces with different exposed surfaces. It is found that friction increases in the order nickel>copper>iron and this ordering is well reproduced by the analytical model.

Finally, as in any good model, it makes various predictions. It suggests that there should be a significant decrease in friction when the material becomes superconducting and also explains the "superlubricity", that is the vanishingly low friction, found for graphite sliding against graphite.

The model also suggests how a thin film on a metal might reduce friction. In this case, a thin layer, even only a few Ångstroms thick, decreases the interaction between dislocations to decrease the drag force. In most cases in tribological research, experimental findings prompt theorists to concoct explanations. Now we appear to have a theory that challenges the experimentalists.

Tribology and Lubrication Technology
June 2007, 59(12) p64

Further Reading:

[1] Merkle, A.P., Marks, L.D. (2007). A Predictive Analytical Friction Model from Basic Theories of Interfaces, Contacts and Dislocations, Tribology Letters, 26, pp. 73–84.

Moving Heat in Nanocontacts

Nanoscale investigations show how roughness affects thermal transport across interfaces

The removal of heat from a sliding interface is an important issue in tribology. Asperity-asperity contact can lead to locally high temperatures that influence additive reactions with surfaces, as well as wear behavior. Surprisingly, though, the way in which heat moves across rough surfaces is remarkably poorly understood. Given that tribological contact occurs between micrometer- or nanometer-scale asperities, these are actually the relevant scales on which to scrutinize heat transfer across an interface. Contact depends crucially on the applied pressure, but experimentally, theoretically, and philosophically, the issue of contact on a very small scale can become problematic, as we have previously discussed in this column [1].

Bernd Gotsmann and Mark Lantz from IBM Research-Zurich in Switzerland have looked at the problem of thermal transmission on the nanoscale, by measuring and modeling heat transport between a scanning silicon tip/cantilever with an integrated resistive heater and a very smooth tetrahedral amorphous carbon (taC) surface [2]. The experiments were carried out in vacuum, so as to eliminate effects of air conductance. The principal quantity of interest to the researchers was the nanoscale pressure dependence of the thermal transfer. During most AFM tip-surface experiments, a pressure dependence of the Hertzian tip-surface nominal contact area is observed, since the tip end can be thought of as being spherical. This makes it difficult to assess the behavior of the contacts on an atomic level, since both pressure and contact area are changing, as load is increased.

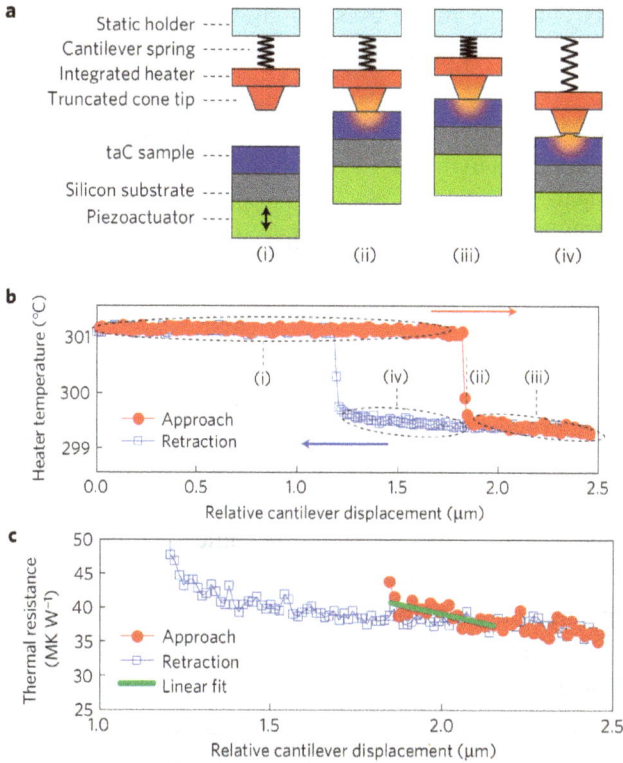

Fig. 1a, Schematic illustrating the different regimes during measurement of thermal conductivity between flat-punch heated AFM tip and taC-coated surface. Initially they are not in contact (i), and the piezo element driving the cantilever will change the distance between them. When in contact (ii–iv), the displacement is translated into a force due to loading and unloading of the cantilever. b, Measured heater temperature as a function of the piezo-displacement (directions indicated by arrows), as the tip is brought into and out of contact with the surface. c, Thermal resistance of the tip–surface contact, calculated from b. The difference between the thermal resistance with the tip in and out of contact with the surface is due to the tip–surface conductance path. Its pressure dependence is shown by the change of resistance as a function of piezo-displacement. This can be linearly fitted (green line). Reprinted by kind permission from Macmillan Publishers Ltd: Nature Materials (Reference [1]), copyright (2013).

Gotsmann and Lantz used a cunning experimental trick to avoid pressure-dependent nominal-contact-area variation: by wearing down the probe tip by sliding it for hundreds of meters over the surface, they could form a conformal, flat-punch arrangement. Thus, pressure-dependent experiments could be relied upon to show the actual effect on thermal conductivity of the number of atoms in contact increasing as a function of load.

The experiment involved increasing the load applied to the surface by the cantilever through the surface-conformal, flattened tip, while simultaneously monitoring the temperature of the integrated heater. Since the only dissipation channel that changed with load was the heat transfer across the tip-surface interface, this measurement could yield the interfacial thermal conductance as a function of applied load (Figure 1).

When contacts are on the nanoscale, the thermal conductivity is quantized, since the diameter of a transport channel is less than the transversal thermal coherence length. Using atomistic simulations, the authors could derive the number of nanocontacts between the flattened tip and the surface, as a function of load. Assigning a quantum of thermal conductance to each nanocontact, they could then calculate the thermal conductance and its pressure dependence. Comparing this and alternative conduction and contact theories with the experimental results, they could show that only by taking atomic roughness and quantized transport into account could the data be accurately predicted. No fitting parameters were used, and physical properties were obtained from the literature. The study thus not only provides insights into the mechanisms of conduction between surfaces with nanoscale roughness, but also provides support for atomistic contact models.

Tribology and Lubrication Technology
August 2014, 70(8) p64

Further Reading:

[1] Tysoe, W.T. and Spencer, N.D. (2009). Contact Conundrum Conquered?, Tribology and Lubrication Technology, 65(6), pp. 88.

[2] Gotsmann, B. and Lantz, M.A. (2013). Quantized thermal transport across contacts of rough surfaces, Nature Materials, 12, pp. 59–65

Topic 3

Lubricating Hard Drives

Despite competition from solid-state information-storage technologies in the form of memory sticks, hard disk drives have managed to keep pace, in terms of cost and storage density, for many decades. This has required a phenomenal industrial R&D effort, and many of the challenges faced have been tribological in nature. We have reported on a number of studies that have been relevant to this extraordinary technological story.

In an effort to reduce the magnetic domain size on hard disks and thereby increase bit density, attempts have been made to locally heat the disk, thus reducing the coercivity and with it the field necessary to perform the write operation. The small gap between write head and rapidly moving disk means that lubrication is essential, but the lubricant film can be affected adversely by the heating, as was discussed in our column entitled *Lubricating Heated Hard Drives*, in February, 2010. The lubricants themselves are typically perfluoropolyethers (PFPEs) that are functionalized with hydroxyl (–OH) groups, which can form hydrogen bonds to the surface as well as chemical interactions with the so-called "dangling bonds" in crevices in the carbon overcoat (*How do Lubricants Really Bond to Hard Drives*, June 2012). The precise way in which such molecules interact with surfaces has also been the subject of sophisticated quantum-chemistry calculations, which have not only explained some of the effects of different PFPE architectures on film properties, but also show potential in the design of better disk-head lubricants (*Computing Chain Conformations*, October, 2011).

Lubricating Heated Hard Drives

Recent experiments show how lubricants are lost from computer hard-disk drives under the influence of laser heating

Since the first computer hard disk drive (see Fig. 1) was introduced in the mid-'50s, there has been an inexorable increase in storage capacity from the first systems that required fifty, two-foot diameter disks to store only 5 Megabytes of memory to today's storage densities of 300 GBit per square inch.

Because the flying height of the slider is equivalent to a jet airliner cruising a few millimeters above the earth, much of this growth in storage density has been due to the development of novel perfluoropolyether lubricants.

However, further increases in memory density will require making smaller and smaller magnetic domains for the magnetic switches that form the bits on the hard disk. Since magnetism is a collective phenomenon, it requires a certain size to form the magnetic domain, so that a smaller domain size means that the resulting magnet becomes weaker.

One solution to this problem has to been to use magnetic materials with a high coercivity, which result in higher magnetic fields at smaller domain sizes. However, switching materials with higher coercivities also require larger magnetic fields and thus is limited by the properties of magnetic materials in the write head.

Since the coercivity decreases with increasing temperature, eventually becoming zero at the Curie temperature, one approach has been to rapidly heat the region of the hard drive where the bit will be written in a technique called thermally assisted magnetic recording. It is anticipated that this approach could lead to magnetic recording densities in excess of 1 TBit per square inch.

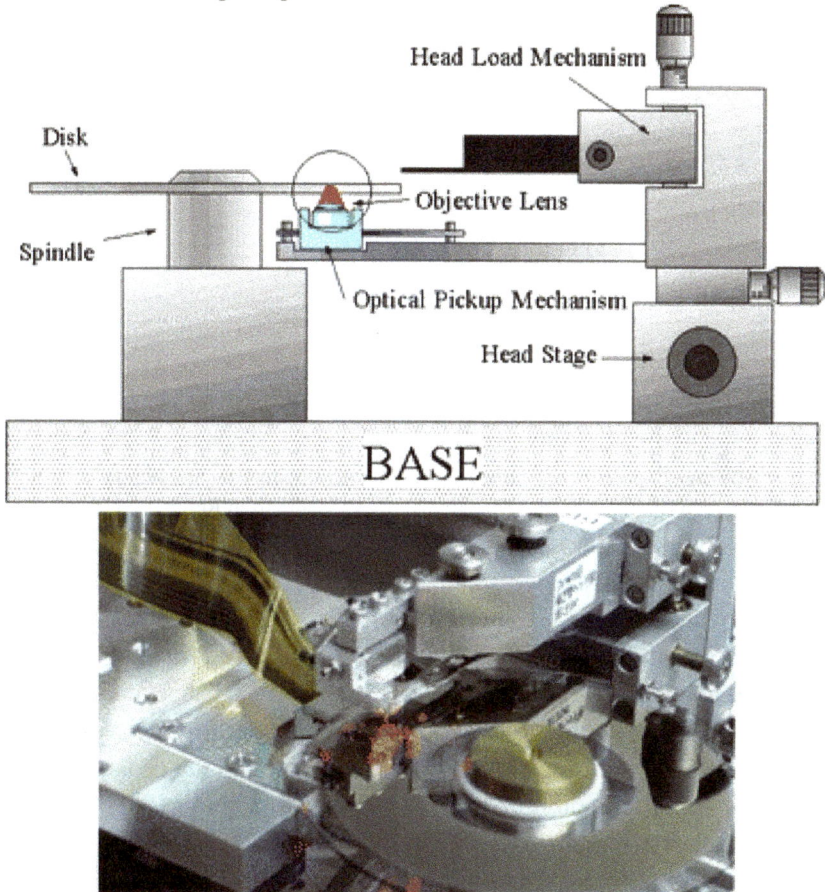

Fig. 1. Hard disk drive and Experimental setup for the thermally assisted magnetic recording system. By kind permission of Springer Science+Business Media from Reference [1].

This process, of course, subjects the lubricant to thermal stresses that can lead to lubricant loss. In order to address this issue, the group of Professor Norio Tagawa at the High Technology Research Center at Kansai University in Japan measured the laser-induced depletion of

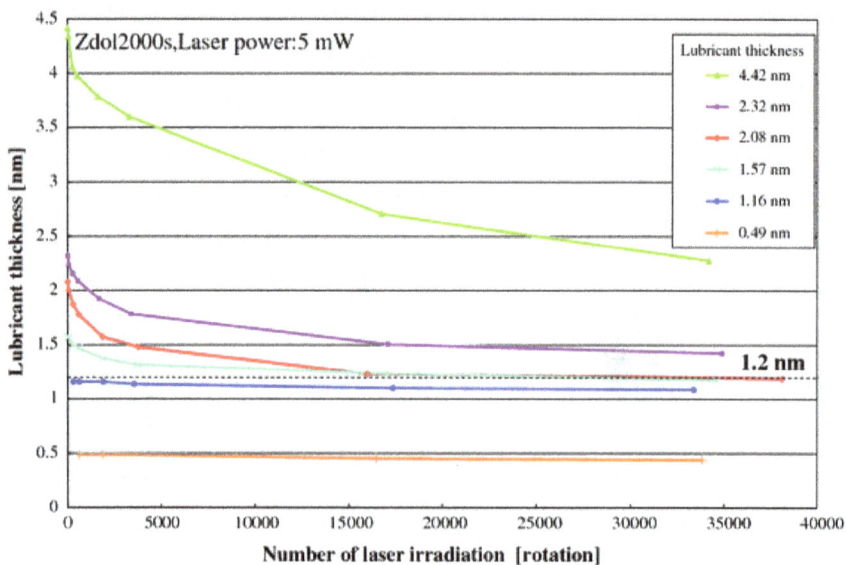

Fig. 2 Relationship between lubricant depletion depth and laser-irradiation duration. By kind permission of Springer Science+Business Media from Reference [1].

various thicknesses of two commonly used hydroxyl-terminated hard disk lubricants, Zdol2000 and Ztetraol2000 [1]. These were deposited onto flat glass slides coated with a thin tantalum film and a carbon overcoat to mimic the disk surface, which were rotated at 8 m/s and illuminated with a 5 mW laser and the resulting film thickness profile was measured. They found that lubricant was lost and the depletion depth increased with radiation time and formed raised ridges on the outer edge of the depleted region. The extent of depletion increased with film thickness but only for films thicker than one monolayer; lubricants bonded directly to the surface were not removed (Fig. 2).

Two possible mechanisms for lubricant loss were identified. The first was lubricant evaporation due to heating and the second was the effect of a decrease in surface tension with increasing temperature that would tend to cause the lubricant to move away from the heated region. They were able to discount the second effect since excess lubricant should accumulate uniformly around the depleted region, while it did not.

In order to explore whether evaporation was responsible for lubricant loss, the surface-temperature rise in the laser beam was simulated and found to result in a maximum temperature increase of ~90 °C. The change in lubricant film thickness when heating to about this temperature was then measured for both lubricants. Lubricant loss was found, and since this temperature is much lower than the lubricant decomposition temperature (~350 °C), this was ascribed to lubricant evaporation.

When this evaporation rate was compared with the rate of lubricant removal rate in the laser beam, almost exactly identical results were found. It seems that solving tribological problems will continue to be at the heart of designing even higher capacity hard drives.

Tribology and Lubrication Technology
February 2010, 66(2) p80

Further Reading:

[1] Tagawa, N., Andoh, H., and Tani, A. (2009). Study on Lubricant Depletion Induced by Laser Heating in Thermally Assisted Magnetic Recording Systems: Effect of Lubricant Thickness and Bonding Ratio, Tribology Letters, 37, pp. 411–418.

How Do Lubricants Really Bond to Hard Drives?

Experiments reveal perfluoropolyether magnetic hard-drive lubricants react with dangling bonds on the surface of crevices in the carbon overcoat

The remarkable properties of magnetic hard disk drives rely on the presence of a thin (1-2 nm) layer of lubricant that is chemically bonded to the surface. These perfluoropolyether (PFPE) lubricants invariably comprise a backbone containing low-surface-energy tetrafluoroethylene oxide ($-CF_2-CF_2-O-$) and difluoromethylene oxide ($-CF_2-O-$) units that provide lubricity. Their hydrophobicity also protects against corrosion. The molecules are terminated at both ends by hydroxyl (-OH) groups that form weak hydrogen bonds with the surface.

Stronger bonds are also formed by a chemical reaction between the hydrogen atoms of the -OH group with so-called "dangling bonds" present on the surface of the carbon overcoat. The hydroxyl hydrogen atoms react with one dangling bond and the resulting reactive terminal oxygen atoms reacts with another to form a strong chemical bond with the surface.

The dangling bonds in the carbon overcoat can be detected using a technique known as electron spin resonance (esr) and it has been found that they disappear when the overcoat is exposed to the lubricant. It has also been found that one type of PFPE lubricant, Z-tetraol, which contains two hydroxyls on each end group, bonds to the carbon overcoat better than Z-dol, which contains only one, confirming the importance of hydroxyl groups in the surface-anchoring reaction.

It would appear, therefore, that lubricant bonding to carbon overcoats on magnetic hard drives is quite well understood. The conundrum, however, is that the dangling bonds on or near the surface also react quite quickly with water vapor or oxygen in the atmosphere. This implies that the dangling bonds required for chemical bonding are present only at a deeper level of the granule composite (Fig. 1).

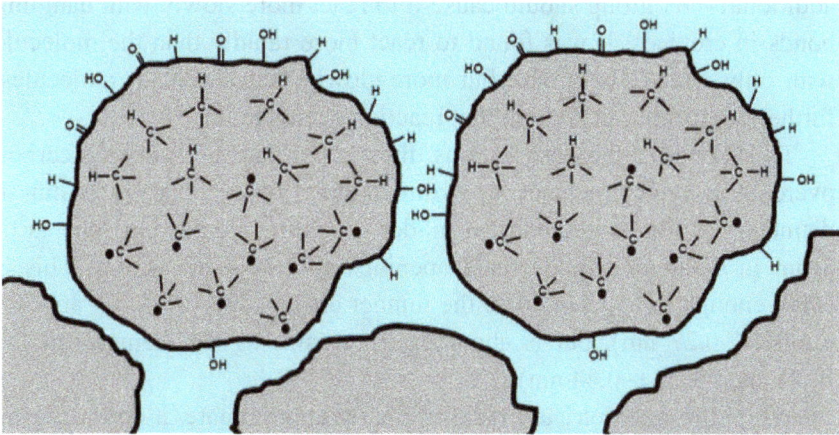

Fig. 1 Sputter-deposited carbon films are aggregates of diamond- and graphite-like granules of several nanometers across. Many dangling bonds remain shielded inside diamond-like granules. By kind permission of Springer Science+Business Media from Reference [1].

In order to test this hypothesis, Drs. Paul Kasai and Tsuyoshi Shimizu of Moresco Corporation in Japan measured the reactivity of a series of PFPE lubricants with similar backbones, but with terminal -OH groups that were tethered *via* -CH$_2$- groups with differing lengths [1]. The diameter of the hydrocarbon tethering groups (~0.27 nm) is less than that of the fluorocarbon backbone (~0.44 nm). Thus, if the thinner tethering groups can move into crevices between granules to reach a dangling bond, longer anchoring groups should be able to access them more easily.

Accordingly, they measured the rate at which the strongly bonded layer formed when using molecules with -OH separated from the PFPE backbone by one, four or six oxygen or -CH$_2$- groups and found that the strongly bonded layer formed the fastest with six groups and the most slowly with one, in accord with their postulate.

They also tested a molecule with five groups, one of which was a — CH(OH)- group, in addition to the terminal -OH group. If all dangling bonds were accessible, this should react more rapidly than molecules with a single -OH terminus, while the steric hindrance afforded by the additional -OH group should cause it to react more slowly with dangling bonds in crevices. It was found to react more rapidly than the molecule with only one -CH$_2$- group, but more slowly than the other molecules, further confirming the notion that reaction occurs in the crevices.

To rationalize the observations, they noted that the sputtered carbon overcoat is typically made up from granules that are about 3 nm in diameter. A simple geometrical model with three contacting spheres of 3 nm in diameter would create openings about 0.5 nm across. This is small enough to accommodate the thinner chains made of -CH$_2$- and -O- groups (~0.27 nm), but is about the same size as the diameter of the PFPE backbone (~0.44 nm).

While this geometrical model is clearly approximate, it does offer an intriguing rationale for their observations. They also note that such a model applies only to thermally cured films and not to those that are cured by ultraviolet light, where the photogenerated electrons interact directly with the PFPE chains, allowing them to bond to the surface.

Tribology and Lubrication Technology
June 2012, 68(6) p96

Further Reading:

[1] Kasai, P.H. and Shimizu, T. (2012). Bonding of Hard Disk Lubricants with OH-Bearing End Groups, Tribology Letters, 46, pp. 43–47.

Computing Chain Conformations

*Quantum mechanics reveals how perfluoropolyether
molecules can lower flying height*

With the advent of faster computers and the ready availability of "parallel" computer codes that allow various parts of a calculation to be carried out simultaneously, it has become feasible to perform quantum-mechanical calculations for physically realistic systems.

This ability has been exploited in many areas of physics and chemistry. For example, quantum mechanics has been used to calculate the elastic properties of materials from the change in energy caused by applying a strain. It is used extensively by chemists to calculate the thermodynamics of reactions and even reaction activation energies.

It is surprising, then, that such methods have not been used to any great extent to understand tribological phenomena. This is, however, being remedied. Quantum calculations were recently used by R. Waltman of Hitachi Global Storage Technologies in San Jose, California to explore the bonding and dynamics of a novel class of lubricants for hard disk drives, consisting of hydroxyl-terminated perfluoropolyethers (PFPE), known as TA-30, invented by D. Shirakawa and colleagues of the Asahi Glass Company in Japan.

The work was motivated by the quest for ever-larger hard-disk-drive storage densities, which requires ever-smaller spacings between the slider element and the disk surface. These have become so small that they are now limited by the thickness and rigidity of the lubricant film, which must also be sufficiently strongly bonded to the disk surface to prevent transfer to the slider. TA-30 achieves this by having hydroxyl groups terminating the PFPE backbone and another hydroxyl-terminated PFPE fragment (a so-called tether) attached to the center of the main chain, forming a Y-shaped molecule.

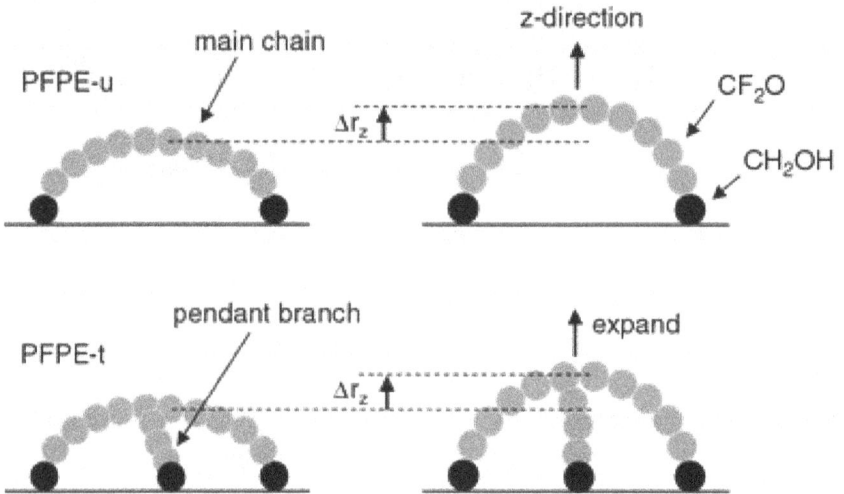

Fig.1 Simple illustration of the computer simulation. By kind permission of Springer Science+Business Media from Reference [1].

Waltman carried out quantum calculations to determine the optimum bonding structure for both a branched molecule analogous to TA-30 and a linear PFPE with hydroxyl groups at either end (Fig. 1). Both molecules were found to bond to the surface *via* their hydroxyl groups with the linear molecule forming a loop, while the branched TA-30 analog also bonded *via* the tether. It was found that the tethered molecule produced a thinner film than the linear one and the resulting structures were in agreement with the experimentally measured monolayer film thickness and polar surface energy.

The molecular rigidity was then explored by calculating the energy change caused upon expanding the molecules by moving the highest part of the loop away from the surface. The energy needed to expand the branched molecule was found to be almost an order of magnitude higher than that for the straight chain (Fig. 2), so that adding the tether substantially improved the molecular rigidity.

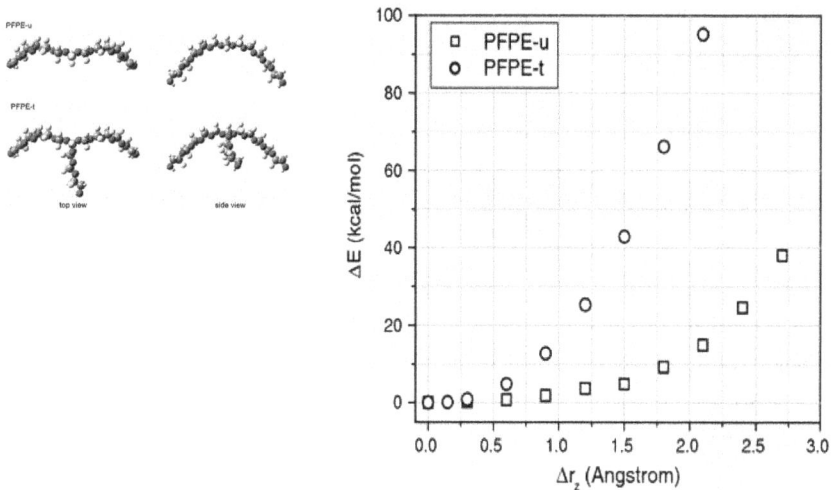

Fig. 2 The total energy, corrected for zero-point energy contributions, as a function of Δr_z for PFPE-u and PFPE-t. By kind permission of Springer Science+Business Media from Reference [1].

By scrutinizing the conformational changes caused by expanding the chains, it was found that the movement in the linear chain was accommodated almost exclusively by rotating around the bonds in the PFPE backbone or by slightly distorting the O-C-O or C-O-C angles. Both of these distortions required much less energy than stretching the bonds themselves, thereby allowing the linear molecule to expand quite easily. In contrast, the additional rigidity imposed by the tethering group on the branched molecule (the TA-30 analog) inhibited such motion, requiring some bonds to stretch. This correspondingly required more energy.

Such quantum calculations on realistic systems can provide insights into the relationship between molecular structure and tribological behavior that will undoubtedly guide our thinking on how to develop new lubricants.

Tribology and Lubrication Technology
October 2011, 67(10) p88

Further Reading:

[1] Waltman, R.J. (2011). Single-Chain Conformational Analysis on the Dynamic Main Chain Expansion of a Tethered Perfluoropolyether Boundary Lubricant Film, Tribology Letters, 43, pp. 175–184.

[2] Shirakawa, D., Ohnishi, K. (2008). A Study on Design and Analysis of New Lubricant for Contact Recording, IEEE Trans. Magn. 44, 3710–3714.

[3] Tagawa, N., and Tani, H. (2010). Conformation and Fundamental Properties of Novel Lubricant TA-30 for Near Contact Magnetic Recording, Tribology Letters, 40, 131–137.

Topic 4

New Materials

One of the holy grails of tribological research is to obtain extremely low friction coefficients, in order to reduce energy consumption, increase fuel efficiency, and generally make it easier to move things around. There is evidence that the ancient Egyptians used lubricants to help them move the massive stones that they needed to build the pyramids. Low friction coefficients can be obtained using hydrodynamic lubrication, but there are many situations in which thick lubricants films cannot be used. The quest for low-friction solid lubricants has often exploited the unique chemical properties of carbon, and this is reflected in the number of *Cutting Edge* articles on new materials involving carbon. This is due to carbon's ability to form a wide variety of chemical structures and a whole field — organic chemistry — is built around its chemical versatility. The discovery of carbon-based materials has also resulted in two recent Nobel prizes, for spherical C_{60} Buckminsterfullerene particles in 1985 and single-layer graphite, graphene, in 2010.

The world record for low-friction materials goes to diamond-like carbon (*Super-slippery solids*, October, 2004) which has a friction coefficient of ~0.008, and spawned a new term, superlubricity.

The discovery of spherical Buckminsterfullerene, or "bucky-balls" created a flurry of excitement that they could perhaps be used as nanoscale ball bearings. It also prompted the synthesis of analogous materials derived from other layered compounds such as molybdenum disulfide. It turned out that the carbon bucky-balls did not retain their structure in the severe environment of a sliding contact. However, other carbon-based nanostructures have been discovered that consist of concentric spheres on top of bucky-balls, dubbed nano-onions or nano-pearls, which do appear to show promise as low-friction lubricant additives (*Novel nano-carbons*, October 2008).

The discovery of monolayer sheets of graphite, graphene, generated considerable interest in their use as lubricants. Unfortunately, as discussed in *A new wrinkle in sliding friction?*, February 2011, the monolayer films were sufficiently flimsy that they tended to form wrinkles as an AFM tip slid over them, leading to high friction. Subsequent attempts were made to make the graphene sheets more rigid by chemically modifying them (*Ironing out the wrinkles in graphene sheets?*, October 2013), which did result in them becoming more rigid. Unfortunately, it appears that stiffer sheets have higher friction, while "ironing out the wrinkles" should, in principle, have lowered it.

Another curious property of conventional graphite, which consists of stacked layers of graphene, is its ability to incorporate other materials between the graphene sheets in a process called intercalation. *In Sliding sandwiches*, October 2012, experiments were discussed in which this trick was used to reduce the interlayer interactions in graphene, in an attempt to produce low-friction films. This trick worked very nicely and produced friction coefficients that were markedly lower than that of untreated graphite.

While carbon-based materials dominated our columns on new materials, the remaining articles on this topic explored other, unusual ways to obtain low-friction interfaces. The conventional wisdom used to be that only materials with three-, four- or six-fold symmetry could fill space. Quasi-crystals have five- or ten-fold symmetry and their discovery led to a Nobel prize in 2011. The frictional properties of these unusual materials were discussed in *Curious quasi-crystals* in February 2006. We eagerly await the announcement of the next Nobel prize for the next wave of novel materials with unusual tribological properties!

In the last of the series on New Materials, entitled *Smoothing the way*, December 2007, we discussed the seemingly paradoxical result that the presence of hard contaminant particles in the lubricant resulted in lower wear of partial hybrid bearings. In spite of many year of investigating new tribological materials, new experimental results continue to surprise us. We can expect many more surprises in the future.

Super-Slippery Solids

The careful design of carbon-based films provides super-low friction

A major goal of much tribological research is to reduce friction between surfaces since this will lead to greater energy efficiency and could potentially result in reductions in wear.

While estimates vary, a relatively large proportion, up to a half, of the energy wasted in an automobile engine, for example, can be attributed to frictional losses. One approach to reducing friction is to coat the surfaces with low-friction films.

A breakthrough in this area came about a decade ago when Ali Erdemir's group at Argonne National Laboratories discovered diamond-like carbon (DLC) films that had friction coefficients as low as 0.008 or less (Fig. 1). These DLC films also had extremely low wear rates, where a one-micron-thick film deposited onto steel endured 17.5 million sliding cycles without wearing through.

The films are deposited from a plasma in a vacuum chamber where the secret to obtaining low-friction films was to ensure that they contained a large amount (more than 40%) of hydrogen. It has been suggested that such a large hydrogen concentration saturates the carbon atoms in the film so that they all become sp^3 hybridized. That is, the bonds around each carbon atom are tetrahedrally distributed in a similar way to diamond except that, while in diamond all carbon atoms are bonded to other carbon atoms, in Erdemir's DLC films, they are also bonded to hydrogen.

Lowering the concentration of hydrogen in the film leads to a larger proportion of the carbon being sp^2 hybridized, where the bonds are arranged in a triangle around each carbon atom (Figure 2). Since now all the electrons on the carbon atoms are not fully occupied by bonding to other atoms, the interaction energy of a surface terminated by sp^2 hybridized carbon is about 5 to 10 times higher than one terminated by sp^3 hybridized carbon.

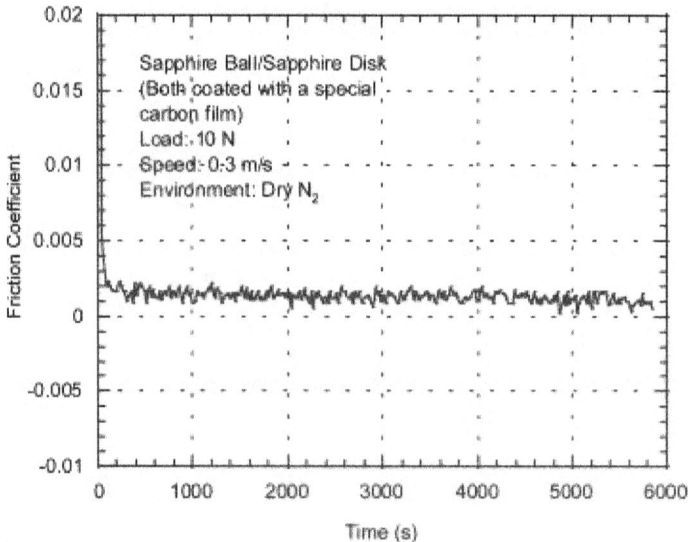

Fig. 1 Frictional performance of a new carbon film providing super-low friction coefficient. Reprinted from Reference [1], with kind permission from Elsevier.

The snag is that such low friction coefficients are only obtained under vacuum or in dry nitrogen, and increase substantially when the film is exposed to water vapor. However, it has recently been shown by Jean-Michel Martin's group in Lyon, France collaborating with scientists in Japan, that DLC films lubricated by a poly alpha-olefin containing 1 wt.% of glycerol monooleate have friction coefficients as low as 0.006 at sliding speeds above 0.1 m/s [2].

This DLC film, unlike the one discovered by the Argonne group, contains almost no hydrogen. Martin analyzed his surface using time-of-flight, secondary-ion mass spectroscopy (ToF SIMS) by bombarding the surface with heavy ions, which remove molecular fragments that are subsequently analyzed mass spectroscopically. New masses were detected corresponding to palmitic acid, as well as significant amounts of hydroxyls ions. These observations showed that the glycerol monooleate had reacted with the DLC film to generate hydroxyl (OH) species on the surface. They speculate that the extremely low friction arises from the low Van der Waals' interaction energy between the two surfaces terminated by the hydroxyl species.

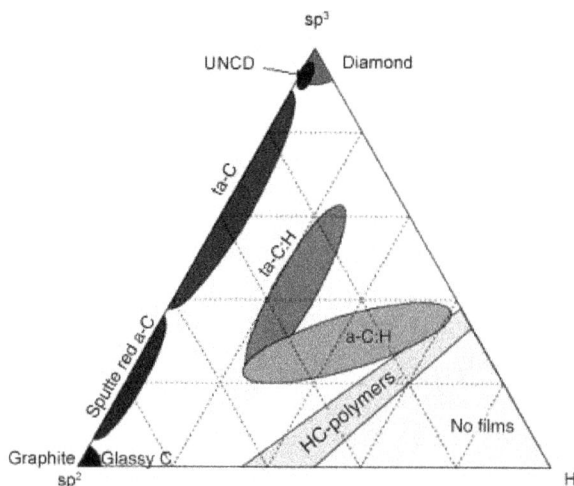

Fig. 2 Illustration of various carbon films with respect to their sp^2- and sp^3-bond characters as well as hydrogen contents. Reprinted from Reference [1], with kind permission from Elsevier.

Tribology and Lubrication Technology
October 2004, 59(12) p64

Further Reading:

[1] Erdemir, A. (2004). Design Criteria for Superlubricity in Carbon Films and Related Microstructures, Tribology International, 37, pp. 577–583.

[2] Kano, M., Yasuda, Y., Okamoto, Y., Mabuchi, Y., Hamada, T., Ueno, T., Ye, J., Konishi, S., Takeshima, S., Martin, J. M., De Barros Bouchet, M. I., and Le Mogne T. (2005). Ultralow friction of DLC in presence of glycerol mono-oleate (GMO), Tribology Letters 18(2), pp 245–251.

Novel Nano-Carbons

Novel, nanoscale carbon materials hold promise for friction and wear reduction

Carbon can exist in many forms, the most common being graphite and diamond. The former has tetrahedrally coordinated carbon, which forms an exceedingly strong bonding network, leading to a very hard material.

Graphite, being a layered compound that has weak Van der Waals' interactions between the hexagonal "graphene" sheets, has long been used as a lubricant. In fact, graphite lubricates better in humid environments due to the intercalation of water between the graphene sheets.

In the mid 1980s, other nanostructures of carbon were discovered, the most famous of these being a nanoparticle containing sixty carbon atoms in a regular arrangement similar to the geodesic domes made famous by Buckminster Fuller and the C_{60} was named buckminsterfullerene in his honor.

Cylindrical versions of buckminsterfulleres, carbon nanotubes, have more recently been synthesized. It was initially thought that these might be promising candidates as solid lubricants since they might act as nanoscale roller bearings, but this turned out not to be the case.

More recently, other carbon nanostructures have been discovered that may hold greater promise. These are nanoscale carbon spheres known as nano-onions and nano-pearls. While both have similar structures, consisting of concentric spheres of graphene sheets, nano-pearls are about 150 nanometers in diameter and have overlapping sheets of graphite flakes that are about 4 nm in size, while the nano-onions, also called giant fullerenes, consist of concentric, spherical fullerene shells (Fig.2).

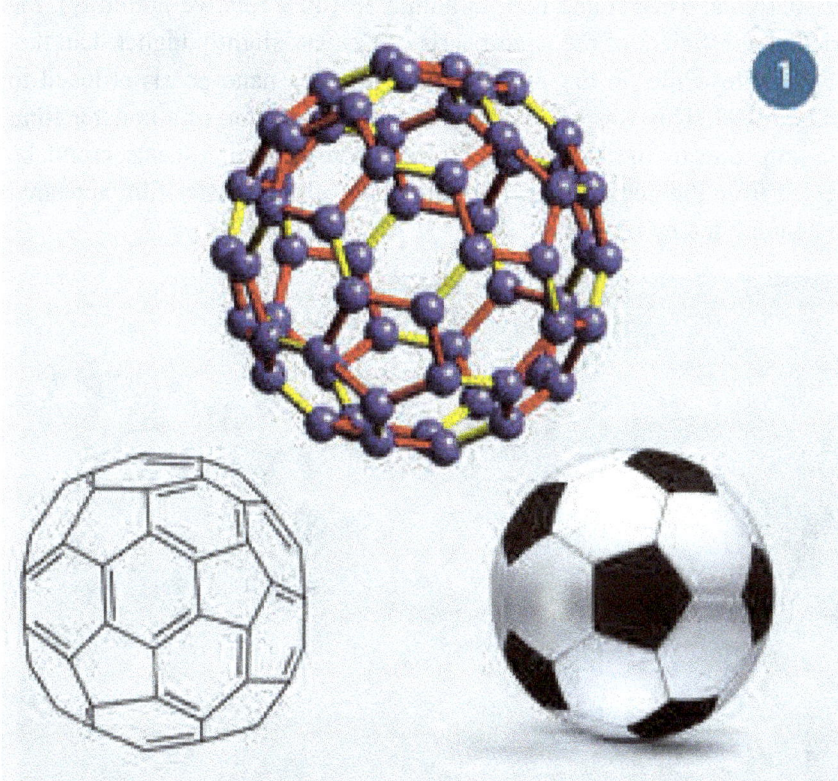

Fig. 1 The structure of buckminsterfullerene in three different representations: ball and stick, a resonance form, and soccer-ball style. Reproduced by kind permission from Reference [1].

Both have interlayer spacings that are larger than found in graphite itself. The tribological properties of nano-pearls were recently investigated by Chad Hunter and co-workers at the Air Force Research Laboratory at the Wright-Patterson Air Force Base in Dayton, Ohio [2].

They argued that the greater interplanar spacing between the graphene sheets in the nano-pearls, along with the presence of some amorphous material between the flakes, might facilitate the shearing of individual layers to generate low-friction films.

They measured the friction of a stainless-steel ball rubbing against silicon that had been loaded with nano-pearls under a Hertzian pressure of ~0.4 GPa. They found that, in humid air (40% relative humidity), the friction coefficient of the nano-pearls (~0.2) was slightly higher than that of graphite while, in dry air, the friction of the nano-pearls reduced to around 0.03. This was partly ascribed to the formation of a transfer film, possibly due to the facility with which the graphene sheets could be peeled from the nano-pearls. The formation of the transfer film appeared to be inhibited by moisture.

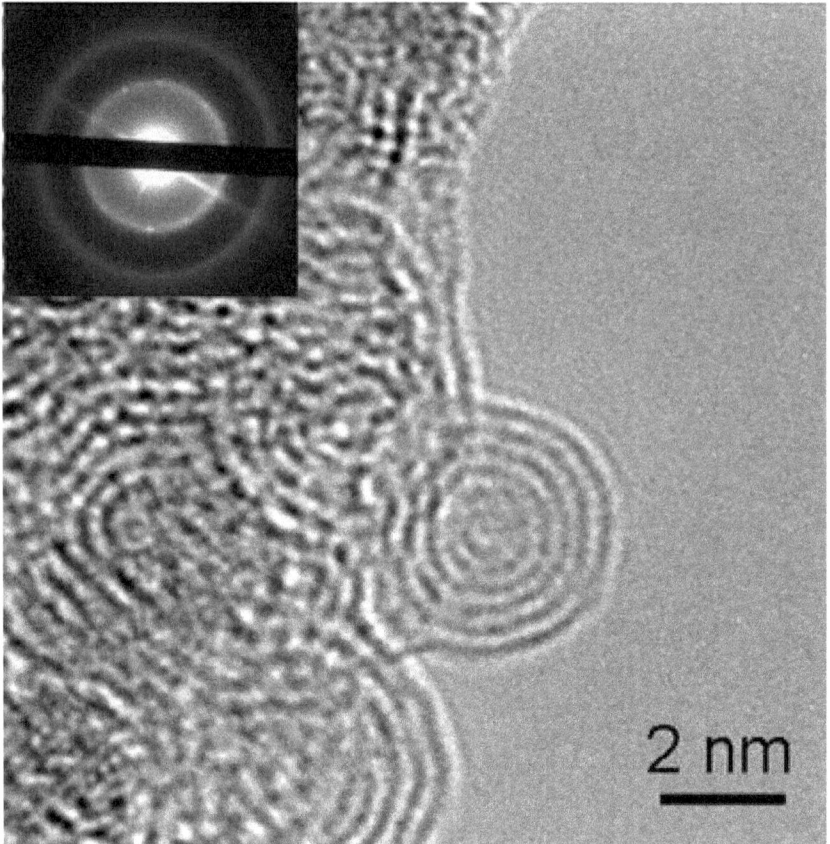

Fig. 2 TEM picture of carbon onions showing the structure of one onion. In frame, the SAED pattern of the whole sample is shown. By kind permission of Springer Science+Business Media from Reference [2].

The friction coefficient of self-mated AISI 52100 steel lubricated by carbon nano-onions dispersed in a poly- alpha-olefin was measured in 30 to 35% relative humidity, also in a pin-on-flat tribometer, in a collaboration between Lucile Joly-Pottuz, Thierry Epicier, Béatrice Vacher and Jean-Michel Martin from the École Centrale de Lyon, France with Nobuo Ohmae from Kobe University in Japan [3].

While the friction coefficient in the presence of the nano-onions at Hertzian pressures between about 0.8 to 1.7 GPa was similar to that measured with graphite, the addition of the nano-onions was found to substantially lower the wear, below that observed with graphite.

While the origin of the reduction in wear is not clear, the authors speculate that it might arise because the nano-onions are softer than graphite or possibly that the surface evolves into a mosaic of graphene sheets and intact nano-onions, bound together by lubricious iron oxide nanoparticles.

It thus seems that, while the initial promise of designing roller bearings based on nanoscale carbon could not be fulfilled, these novel materials nevertheless appear to hold promise, both for friction and wear reduction.

Tribology and Lubrication Technology
October 2008, 64(10) p72

Further Reading:

[1] Harrison, P. and McCaw, C. (2011). Symmetry of Buckminsterfullerene, Education in Chemistry, June 2011, pp. 112–113.

[2] Hunter, C.N., Check, M.H., Hager, C.H. Voevodin, A.A. (2008). Tribological Properties of Carbon Nanopearls Synthesized by Nickel-Catalyzed Chemical Vapor Deposition, Tribology Letters, 30, pp. 169–176.

[3] Joly-Pottuz, L., Vacher, B., Ohmae, N., Martin, J.M., Epicier, T. (2008). Anti-wear and Friction Reducing Mechanism of Carbon Nano-onions as Lubricant Additives, Tribology Letters, 30, pp. 69–80.

A New Wrinkle in Sliding Friction

The friction of an atomically thin layer is found to be higher than that of multiple layers due to puckering of the thinner films during sliding

Layered compounds, such a graphite and molybdenum disulfide (MoS_2), have been used as solid lubricants for many years. Recently it has been discovered that a single, atomically thick sheet of graphite, dubbed graphene, has properties that are quite distinct from those of the bulk material. It has an extremely high electrical conductivity, is remarkably strong and, in spite of its thinness, is impervious to gases.

Graphene has a myriad of potential technological applications from transistors that are a few nanometers across to efficient and cheap photovoltaics. The fundamental and technological importance of graphene was recognized in this year's Nobel prize in physics, awarded to Andre Geim and Konstantin Novoselov from the University of Manchester in the UK, who discovered that graphene could be easily prepared by "exfoliating" graphite. That is, single graphene layers could be pulled from bulk graphite just using a piece of adhesive tape.

In view of the unusual properties found for graphene, Prof. Rob Carpick and his group from the University of Pennsylvania, along with Prof. Jim Hone of Columbia University and collaborators, wondered how the friction of a single layer of a layered compound would differ from that of the bulk material. They used atomic force microscopy (AFM) to explore this question for several layered compounds with different electrical properties, graphene (a semimetal), molybdenum disulfide (a semiconductor), hexagonal boron nitride (an insulator) and niobium diselenide (a metal), all deposited onto an oxidized silicon substrate (Fig.1). The first three have been used as solid lubricants.

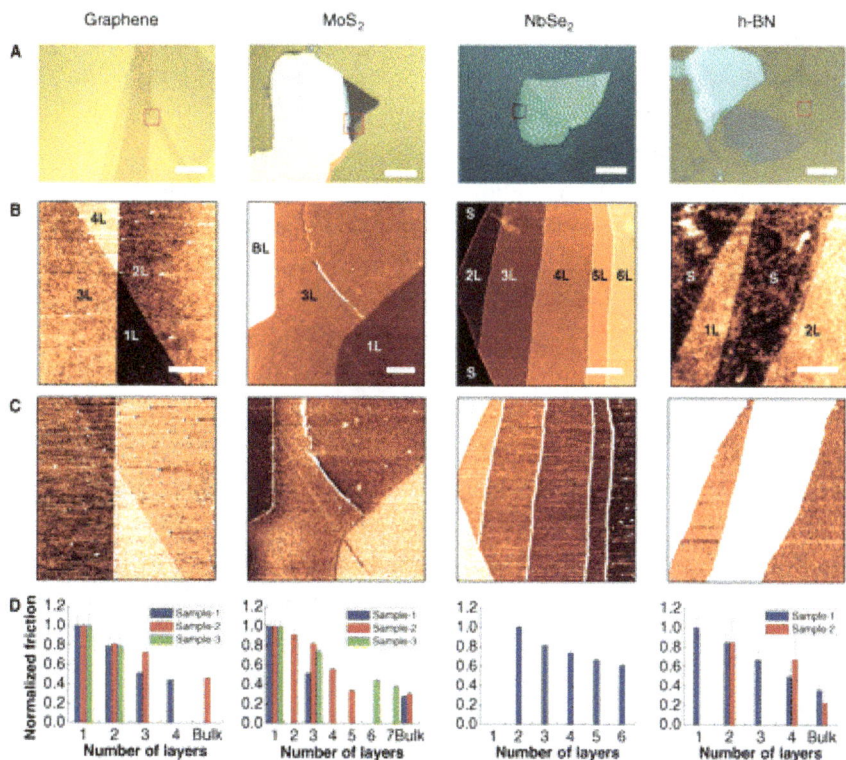

Fig. 1 Optical and AFM images of atomically thin sheets of (from left to right) graphene, MoS_2, $NbSe_2$, and h-BN on silicon oxide. (**A**) Bright-field optical microscope images of thin sample flakes. The red dotted squares represent subsequent AFM scan areas. Scale bars, 10 μm. (**B** and **C**) Topographic and friction (forward scan) images measured simultaneously by AFM from the indicated areas. 1L, 2L, 3L, etc. indicate sheets with thicknesses of one, two, three, etc., atomic layers. BL ("bulk-like") denotes an area with a very thick flake, and S represents an area with bare SiO_2 substrate. Scale bars, 1 μm. (**D**) Friction on areas with different layer thicknesses. For each sample, friction is normalized to the value obtained for the thinnest layer. Error bars represent the standard deviation of the friction signals of each area. In each chart, the same color represents data from the same sample. From Reference [1]. Reprinted with kind permission from AAAS.

AFM is the ideal tool for such experiments, since the number of layers in the film can be measured by imaging the sample in contact mode, and the friction can measured from the torsion of the cantilever during sliding. In all case, a single atomic layer was found to have higher friction than multiple layers and friction decreased to the bulk value for a film of about four to five layers thick.

The group carried out a number of control experiments to identify possible origins for the variation in friction. Since all materials showed the same trend, their electronic properties were apparently not important in causing the friction effect. They also excluded the effects of load, sliding speed, tip material and humidity. The effect of the support was excluded by performing similar experiments for sheets suspended over holes in the substrate, but the same trend was still observed. The effect was entirely due to the thickness of the film.

The origin of the observed behavior was identified by carrying out simulations of a tip sliding across a flexible membrane. Carpick and colleagues discovered that for the thinnest (monolayer) films, a wrinkle forms during sliding at the leading edge of the tip. This disappears as the film becomes thicker. It is suggested that the puckering of the thin substrate results in a higher contact area between the tip and substrate, resulting in the higher friction forces observed experimentally.

The authors suggest that these results can lead to the rational design for such monolayer materials in nanomechanical applications, by ensuring that the monolayer sheet is strongly enough attached to the substrate to prevent it from buckling.

Tribology and Lubrication Technology
February 2011, 67(2) p72

Further Reading:

[1] Lee, C., Li, Q., Kalb, W., Liu, X.-Z., Berger, H., Carpick, R.W. and Hone, J. (2010). Frictional Characteristics of Atomically Thin Sheets, Science, 328, pp. 76–80.

Ironing Out the Wrinkles in
Graphene Sheets

Chemically modifying graphene changes its frictional properties

A previous Cutting Edge article discussed the unusual nanotribological properties of atomically thin films of graphite, known as graphene, on surfaces [1]. It was found that the high friction of a single graphene layer, as measured in an atomic force microscope (AFM), could be traced to the formation of wrinkles at the leading edge of the tip, thereby increasing the contact area between the tip and film. These observations suggested that it should be possible to modify the properties of graphene, in order to tune its tribological properties. Stiffer films should presumably result in lower friction.

The group of Professor Jeong Young Park at the Center for Nanomaterials and Chemical Reactions at the Korea Advanced Institute of Science and Technology along with collaborators at Konkuk University took advantage of the ability to chemically functionalize graphene sheets by reaction with fluorine, hydrogen or hydroxyl groups [2]. This changes the number of atoms to which each carbon is bonded, thus modifying the mechanical properties. In pristine graphene, each carbon atom is bonded to three other carbon atoms, resulting in the well-known honeycomb structure. Reaction with another atom leads to carbon atoms being bonded to four other atoms, causing the structure around each carbon atom to become closer to tetrahedral.

Fluorinated graphene was synthesized by reacting a chemical vapor deposition-grown graphene layer with xenon hexafluoride. Subsequent elemental analysis using X-ray photoelectron spectroscopy showed that this resulted in a material with a chemical formula C_4F. Functionalization by hydrogen or hydroxide species was achieved by applying a positive or negative bias to an AFM tip in the presence of water vapor. The high electric field between the tip and surface caused the water to dissociate. Application of a positive bias caused the oxidation of the graphene, while a negative bias produced a narrow strip of hydrogen-modified graphene. The strip is surrounded by pristine graphene, thus allowing the friction of the modified and unmodified graphene to be compared in a single experiment. The resulting structural modifications were confirmed by means of Raman spectroscopy.

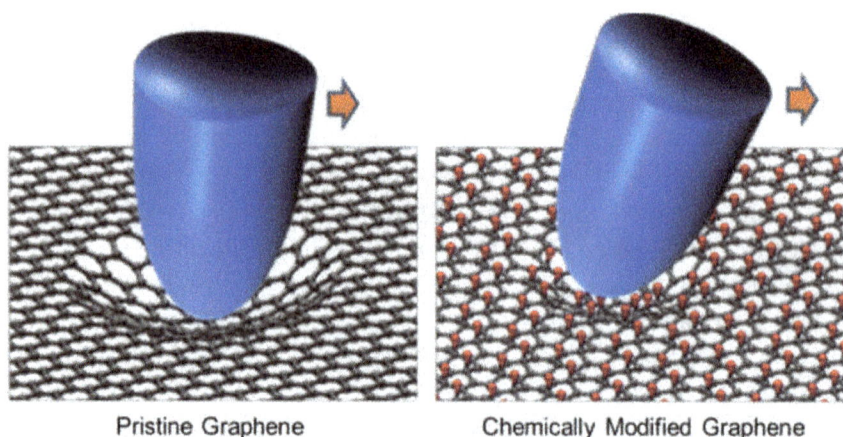

Pristine Graphene   Chemically Modified Graphene

Fig. 1 Illustration of nanoscale friction measurements on pristine graphene and chemically modified graphene. By kind permission of Springer Science+Business Media from Reference [2].

The nanoscale friction of the modified and pristine graphene were compared and revealed that the fluorine-modified surface had a friction force about six times higher than that of pure graphene (Fig. 1). Oxygen modification caused the friction to increase by a factor of seven, and hydrogen by a factor of two. However, the adhesion between the tip and the modified surfaces was identical to that on pristine graphene.

In order to determine the way in which chemical modification had influenced the mechanical properties, the researchers carried out first-principles quantum calculations using density functional theory (Fig. 2). To estimate the in-plane stiffness, they calculated the change in energy of the sheets while changing the spacing between the carbon atoms. The energy was found to vary parabolically with a change in spacing, enabling the force constant to be calculated. A modest reduction in the in-plane force constant by ~35% was found.

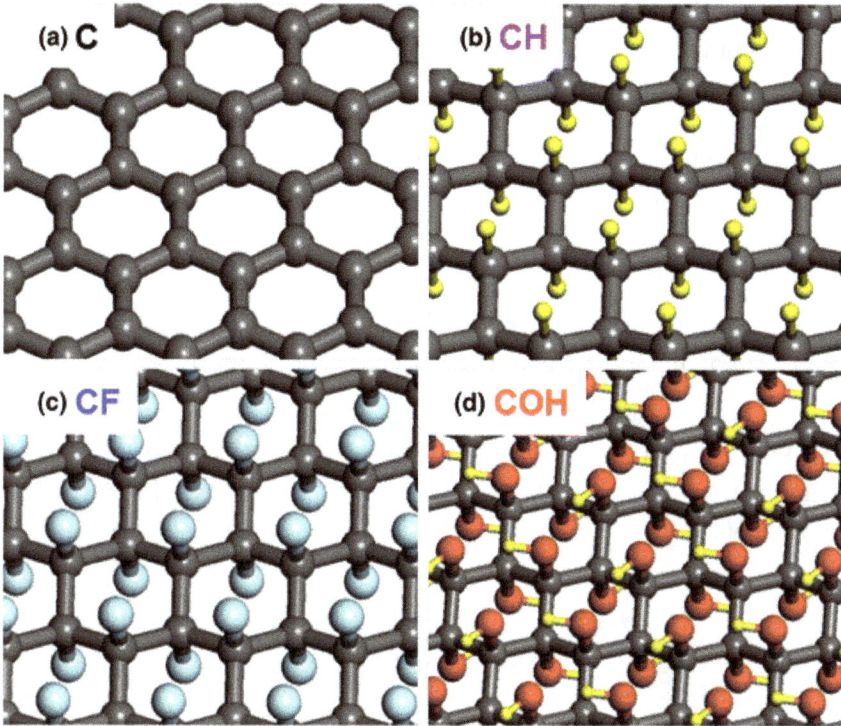

Fig. 2: Atomic models of (a) pristine graphene (C), (b) hydrogenated graphene (CH), (c) fluorinated graphene (CF), and (d) hydroxidized graphene (COH). By kind permission of Springer Science+Business Media from Reference [2].

The out-of-plane stiffness was obtained in a similar way, by calculating the energy of the film after moving a carbon atom along a direction perpendicular to the plane. The energy similarly showed a quadratic dependence on displacement. This revealed a substantial effect of chemical modification, with the hydrogenated film being ~1.4 times stiffer than pure graphene, and fluorine increased the out-of-plane stiffness by a factor of ~5.5 and oxidation by a factor of ~7.

The increase in out-of-plane stiffness correlates very well with the relative friction coefficient; the stiffer the sheets, the higher the friction. However, stiffer films should wrinkle less and it was suggested that bending vibrations could be the main source of frictional energy dissipation. Clearly, chemical modification of graphene does influence friction, but the cause still remains elusive, suggesting the need for more detailed theoretical studies.

Tribology and Lubrication Technology
October 2013, 96(10) p96

Further Reading:

[1] Tysoe, W.T, and Spencer, N.D. (2011). A New Wrinkle in Sliding Friction? Tribology and Lubrication Technology 67, pp. 72.

[2] Ko, J.H., Kwon, S., Byun, I.-K., Choi, J.S., Park, B.H., Kim Y.-H. and Park, J.Y. (2013). Nanotrobological Properties of Fluorinated, Hydrogenated and Oxidized Graphenes. Tribology Letters, 50, pp. 137–144.

Sliding Sandwiches

Intercalating metal halides reduces
inter-plane interactions and lowers friction

Lamellar compounds consist of strongly bonded planar layers that have weak van der Waals' forces between them. The presence of the weak interlayer interactions has been suggested to facilitate shear along directions parallel to the layers to lower the friction coefficient.

There are several well-known examples of such lubricious lamellar compounds, the best known of which are graphite and molybdenum disulfide. A product of tribochemical reactions with chlorine-containing additives, ferrous chloride ($FeCl_2$) is also a layered compound and has low friction.

The weak interlayer forces in lamellar compounds also result in unique properties by allowing chemicals to be "intercalated" between these layers. Graphite, in particular can intercalate a wide variety of materials, including alkali metals and this property is exploited to store lithium in lithium-ion batteries that are commonly used in laptop computers, for example.

Since the resulting graphite intercalation compounds also have lamellar structures, they are also expected to display low friction. Since the presence of the intercalated layer separates the graphene (single graphite) sheets, the interlayer interaction energy might be expected to be lower than for graphite, and the intercalation compounds might therefore be expected to have lower friction.

Unfortunately, it is difficult to measure the interaction energy directly. To solve this problem, Drs. Jean-Louis Mansot and Philippe

Thomas of the Université des Antilles et de la Guyane, in Guadeloupe compared the results of quantum-theory calculations of the structures of intercalation compounds with their frictional behavior [1].

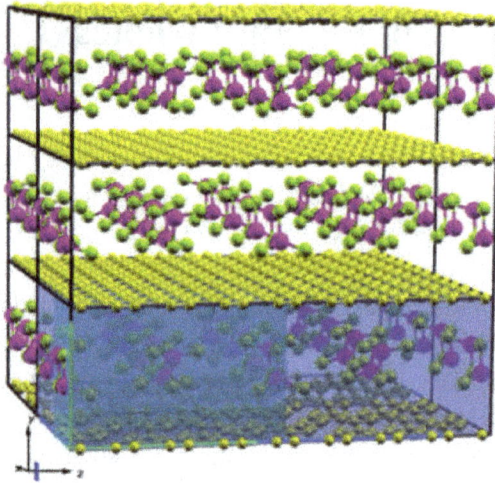

Bulk (3D periodic system) with E_{Bulk} as the total energy of the primitive cell material

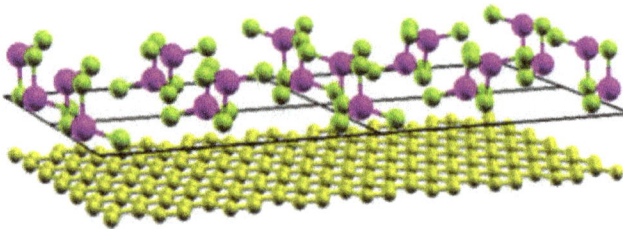

Isolated layer (2D periodic system deduced from the bulk) with $E_{Isolated\ layer}$ as the total energy

of the primitive cell of the isolated layer.

$$IE = E_{Bulk} - E_{Isolated\ layer}$$

Fig. 1 Depiction of the structure of $Al_4Cl_{12}C_{48}$. The interaction energy between layers is obtained via the following formula: $I_E = E_{Bulk} - E_{Isolatedlayer}$ where E_{Bulk} is the total energy of the primitive cell in the bulk material and $E_{Isolated\ layer}$ is the total energy of the primitive cell in the isolated layer. By kind permission of Springer Science+Business Media from Reference [1].

They examined the behavior of several metal-halide intercalation compounds synthesized with ferric chloride ($FeCl_3$), aluminum chloride ($AlCl_3$) and antimony pentachloride ($SbCl_5$). They are all known to form intercalation compounds, in which the intercalated molecules occupy all of the spaces between the graphene layers. This results in an expansion of the interlayer spacing of pure graphite (0.335 nm) to ~0.95 nm for the intercalation compounds. All of the metal-halide intercalation compounds have a similar, repeating graphene-chlorine-metal-chlorine structure.

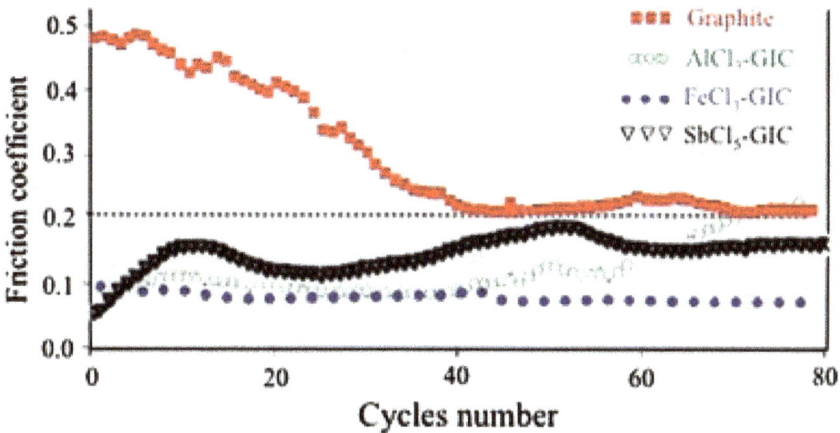

Fig. 2 Friction-coefficient evolution recorded under high-purity Ar atmosphere for the reference compound (graphite) and the three graphite intercalation compounds. By kind permission of Springer Science+Business Media from Reference [1].

In order to calculate the interlayer interaction energy, Drs. Mansot and Thomas used density functional theory to calculate the total energy of a unit cell of the intercalation compound, which includes the interaction between the layers, and compared this with the energy of an isolated layer (Fig. 1). The difference between the two energies is a measure of the interlayer interaction. The interaction energy between graphene layers in pure graphite (~4.3 kJ/mol) was significantly reduced in the intercalation compounds to ~0.1 kJ/mol. By analyzing the results of the electronic structure calculations, the reduction was traced to both an increase in the distance between the graphene layers and the very weak graphene-intercalant interactions.

This suggests that the intercalation compounds should have lower friction than pure graphite and this was tested by comparing the friction coefficient of graphite and the intercalation compounds under a high-purity argon atmosphere, where the friction coefficient of intercalation compounds (~0.09) was indeed significantly lower than that of pure graphite (0.22) (Fig. 2). In the case of the intercalation compound with ferric chloride, the friction coefficient remained low for 80 rubbing cycles, while with the other intercalation compounds the friction coefficient slowly rose to that of pure graphite with continued rubbing. This was explored by using Raman spectroscopy to examine the structure of the antimony pentachloride intercalation compound after various numbers of rubbing cycles.

Indeed it was found that the partial de-intercalation occurred after only 10 cycles and was complete after 80, rationalizing the observed change in friction. Such a close interplay between theory — in this case, first-principles quantum calculations — and experiment serves not only to test our chemical intuition but also provides insights into the behavior of the materials.

Tribology and Lubrication Technology
October 2012, 68(10) p88

Further Reading:

[1] Delbé, K., Mansot, J.-L., Thomas, Ph., Baranek, Ph., Boucher, F., Vangelisti, R. and Billaud, D. (2012). Contribution to the Understanding of Tribological Properties of Graphite Intercalation Compounds with Metal Chloride, Tribology Letters, 47, pp. 367–379.

Curious Quasi-Crystals

Atomic force microscopy measurements reveal high friction anisotropies for sliding on quasi-crystal surfaces

In previous columns, we have discussed the effect of commensurability on sliding [1]. A more fundamental question is whether the existence of periodicity itself is important in friction. In principle, this notion should be relatively easy to test. One would merely have to prepare a material in its crystalline and amorphous forms and compare their frictional behavior. The problem with this is that the crystalline and amorphous forms are also likely to be chemically different, thereby potentially masking any effects that may be due to structural differences.

A solution to this problem has recently been provided by work from the groups of Pat Thiel at the Ames Laboratory and Miquel Salmeron, with post-doctoral researcher Jeong Park, of the Lawrence Berkeley National Laboratory, who examined the frictional anisotropy of quasicrystalline materials [2].

These are dubbed "quasi"-crystals since they possess 5- or 10-fold symmetry (Fig. 1). Strictly speaking, crystals with 5- or 10-fold symmetry are impossible; they do not form an allowed, so-called Bravais lattice. The argument for this is straightforward; all of the unit cells of a crystalline material have to fit together to completely fill space. Thus, while it is possible to completely cover a surface with triangular (3-fold), square (4-fold) or hexagonal (6-fold) tiles without any gaps, this simply cannot be done with 5-fold surfaces.

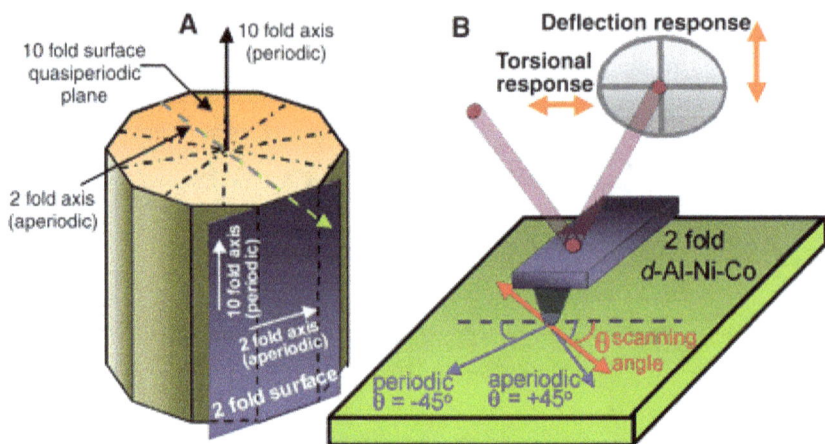

Fig. 1 (A) Schematic model of a decagonal Al-Ni-Co quasicrystal, showing the orientation of decagonal and twofold planes. The 2-fold plane is periodic along the 10-fold direction and a periodic along the 2-fold direction. (B) Schematic of the cantilever and the scanning geometry during friction studies. From Reference [2]. Reprinted with kind permission from AAAS.

The way nature allows such apparently forbidden materials to exist is to have only local 5-fold rotation symmetry but no translational symmetry. The distances between atoms however are not random, but follow patterns that conform to the curious sequences called Fibonacci series. A particular class of these materials, the decagonal ones, have both periodic and aperiodic directions in the lattice made of planes with 5-fold symmetry stacked in a periodic fashion.

The Thiel and Salmeron groups exploited this effect by cutting a decagonal Al-Ni-Co alloy quasicrystal perpendicularly to its 10-fold rotational axis. The surface obtained in this manner has both periodic and aperiodic ordering along different directions that are separated by 90°.

The sample was cleaned in ultrahigh vacuum, and was studied using a combined scanning tunneling/atomic force microscope (STM/AFM) apparatus (Fig. 1). STM is capable of obtaining atomic-resolution images of the surface to allow the periodic and aperiodic directions to be identified. AFM was then used to measure the frictional force *versus* normal force along the periodic and aperiodic directions for effective loads below about 200 nN.

It is crucial that the surface deform elastically in spite of the small contact radius of the AFM tip (~10 nm) to ensure that the original surface structure is not destroyed during sliding. In order to achieve this, the adhesion force between the tips and quasicrystal surface was decreased by passivating the tip with a molecular layer of hexadecane thiol.

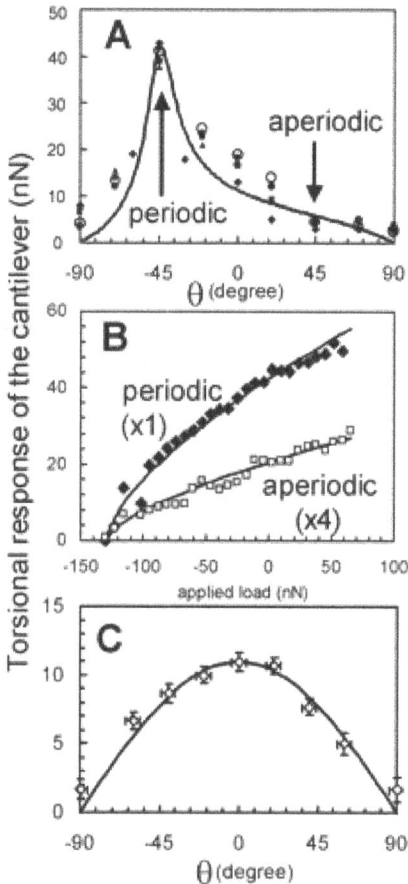

Fig. 2 (A) Torsional response of the cantilever as a function of scanning angle on the twofold surface of the Al-Ni-Co decagonal quasicrystal. The solid line shows the calculated torsional response with scanning angle for an anisotropy factor (ratio of friction forces) of 8. (B) Torsional response as a function of applied load in both periodic and aperiodic directions. The ratio of shear stress in each direction derived from the fits is 8.2 ± 0.4. (C) Plot of the torsional response as a function of scanning angle on an isotropic silicon oxide surface with a roughness of <0.3 nm. The solid line shows the calculated torsional response. From Reference [2]. Reprinted with kind permission from AAAS.

AFM experiments showed that the friction force while sliding along the periodic direction was about eight times larger than when sliding along the aperiodic direction; friction does apparently depend on periodicity (Fig. 2). In order to check that this effect was not due to experimental artifacts, they measured the frictional anisotropy of amorphous silicon oxide and an aluminum oxide layer formed by air-oxidation of the quasicrystal surface, and found no dependence on sliding direction. Clearly, this anisotropy is not due to commensurability since the two surfaces are completely different.

The authors concluded that the differences could be due to dissipation either by electronic or vibrational excitations. As the authors note "Our results call for a detailed modeling of the generation and propagation mode of electronic and phonon (vibrational) excitations". Over to the theorists for the next chapter in this saga!

Tribology and Lubrication Technology
February 2006, 62(2) p64

Further Reading:

[1] Tysoe, W.T, and Spencer, N.D. (2005). The Continuing Contact Conundrum, Tribology and Lubrication Technology 61(10), pp. 72.

[2] Park, J.Y., Ogletree, D.F., Salmeron, M., Ribiero, R.A., Canfield, P.C., Jenks, C.J., and Thiel, P.A. (2005). High Friction Anisotropy of Periodic and Aperiodic Directions on a Quasicrystal Surface, Science, 309, pp.1354–1356

Smoothing the Way

Dirt in bearings leads to dents, leads to stress, leads to rolling contact fatigue. Ceramic hybrid bearings may be the answer.

The presence of contaminated lubricant in bearings is an unpleasant fact of tribological life. Whether it be the result of negligent service or an inevitable consequence of dusty operating conditions, dirt happens, and it can cause a significant reduction in the bearing lifetime. One reason for this appears to be the presence of protrusions, or shoulders, around dents made by the dirt particles, leading to high local stress concentrations that increase the danger of rolling contact fatigue.

The use of ceramic hybrid bearings, which consist of ceramic rolling elements in a steel raceway, may be a solution to problems arising from dirty lube. While these bearings have longer rolling contact fatigue lifetimes than conventional steel bearings at high speed, their performance at lower speeds is not as impressive. There have been indications, however, that the presence of dirt changes the picture entirely. Reports from both industry and academia have suggested that in a contaminated lube, ceramic hybrid bearings seem to have longer lifetimes than conventional steel bearings, even at lower speeds. It has been speculated that this is due to a smoothing effect of the hard, ceramic balls on the damaged steel raceway surfaces.

A team under the leadership of Patrick Wong at the City University of Hong Kong has, in a recent article in *Tribology Letters,* brought closure to the earlier work by a nice combination of bearing testing, tailored experiments, microscopy, and finite-element modeling [1]. First, they deliberately made dents in a steel disc, and then rolled it in an oil bath against discs made of either steel or silicon nitride — a hard ceramic. The flattening of the shoulder around the dent was clearly more effective in the case of the ceramic counter-surface. Next, a finite-element model, based on input data from the rolling test, was constructed to determine the effect of the protrusions on the stress distribution in the dent region: this is critical in determining the rolling contact fatigue lifetime. It could then be calculated, from the Lundberg-Palmgren model for bearing fatigue life, that the dent in the case of the silicon nitride/steel pairing would be six times less damaging than the steel-steel pairing, in terms of rolling contact fatigue.

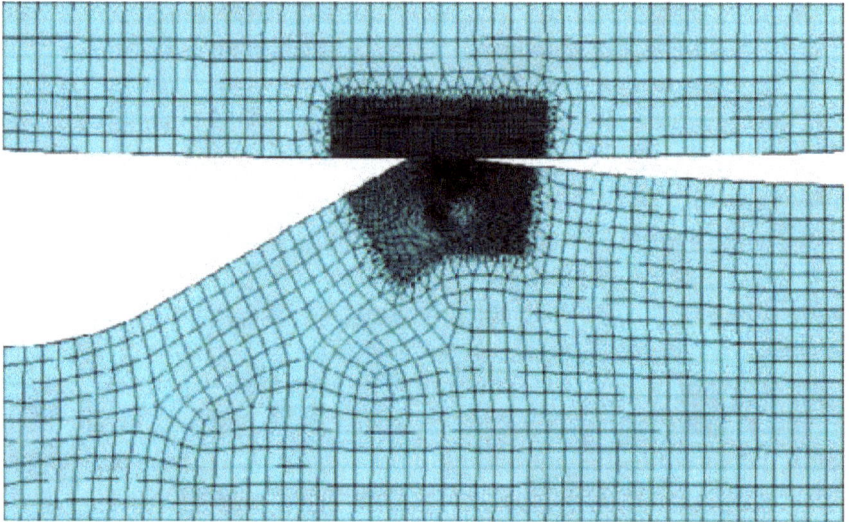

Fig. 1 Finite-element meshing of the dent area on a ceramic bearing. The area around the protrusion uses denser mesh to enhance the accuracy of the calculation. By kind permission of Springer Science+Business Media from Reference [1].

Finally, a real bearing test was carried out, in which a conventional steel bearing was compared with a bearing in which two of the nine steel rolling elements had been replaced by silicon nitride balls. This "partial hybrid bearing" was placed on a test rig, on the same shaft as an equivalent bearing consisting of steel elements only, and the entire system lubricated with oil that had been "spiked" with unpleasant-looking silicon carbide particles. The partial hybrid bearings showed less wear than the steel bearings, and electron microscopy of the steel raceway surfaces following the test showed less severe wear scars and smaller dents in the partial hybrid bearing.

Tribology and Lubrication Technology
December 2007, 63(12) p64

Further Reading:

[1] Wong, P.L., He, F. and Wan, G.T.Y. (2007).Experimental Study of the Smoothing Effect of a Ceramic Rolling Element on a Bearing Raceway in Contaminated Lubrication, Tribology Letters, 28, pp. 89–97.

[2] Wan, G.T.Y, Gabelli, A. and Ioannides, E. (1997). Increased Performance of Hybrid Bearings with Silicon Nitride Balls, Tribology Transactions, 40, pp. 701–707.

[3] Kahlman, L. and Hutchings, I.M. (1999). Effect of Particulate Contamination in Grease-Lubricated Hybrid Rolling Bearings, Tribology Transactions, 42, pp. 842–850.

Topic 5

New Methods

Tribology is no exception to the observation that many breakthroughs in science involve inventions in instrumentation and methods. We have written about a number of innovations in our column, that we believe have the potential to move our field forward.

Machining is a key technology in the industrial world, and tribology plays an important role, both in terms of the friction, and therefore heat, that is generated between cutting tool and workpiece, and in the lifetimes of cutting tools, which are limited by wear processes. Lubrication is clearly crucial in this context. We covered a novel *in situ* approach to monitoring the lubricant behavior at the cutting-tool-workpiece interface, in which a fluorescent lubricant is observed through a transparent cutting tool (Monitoring the Cutting Edge, October, 2007). This shows promise as a technique for providing a window into this normally buried and difficult to access tribological interface.

If friction is challenging to understand on a fundamental level, wear is a phenomenon that lies at an even higher level of complexity. Progress in fundamentally understanding wear has been slow, although recent methods involving tribological experiments in which the wearing interface has been examined directly using a transmission electron microscope (TEM) have been promising. In Tribochemical Wear in a Transmission Electron Microscope, February, 2013, we featured an article describing how researchers had studied the rubbing of a tungsten tip against a free-standing diamond-like carbon film in a TEM, measuring the thinning of the sample due to wear by increased electron transmission and resulting changes in the hybridization of surface carbon by electron-energy loss measurements. TEM was also applied to another

137

problem in the area of solid lubrication, namely that of the behavior of molybdenum disulfide nanoparticle in sliding contacts (Imaging Rolling Nanoparticles, June 2011). TEM revealed that the MoS_2 nanoparticles undergo rolling, sliding, and exfoliation, depending on the tribological conditions applied.

Wear measurements, when the actual quantity of wear is very small, are notoriously difficult. One approach that has been widely used in the automobile industry in Europe is to irradiate the sliding interfaces in a cyclotron, followed by sensitive measurements of radioactivity in the oil. This has some obvious difficulties and disadvantages. A new approach that does not involve radioactivity (The Gold Standard for Wear, February, 2009) has been suggested, in which gold atoms are first implanted into the test surfaces, and Rutherford backscattering is used to measure the depth distribution before and after the test, yielding precise numbers for the amount of material removed.

The use of focused ion beams (FIB) is a relatively new approach to sculpting materials on a sub-micrometer scale, either to extract subsurface layers for subsequent microscopy, or to produce tiny test devices for exploring phenomena relevant to microelectromechanical systems (MEMS). A fundamental study involving the sliding contact between a titanium sphere and a series of FIB-fabricated, micrometer-scale grooves in a silicon wafer (Stuck in a Rut, August, 2011) revealed that friction was at its maximum at exactly the point where the sphere was simultaneously touching both walls and base of the grooves.

While most MEMS devices are fabricated from silicon, we also reported on an attempt to use the photoresist SU-8 to make microdevices that involve a tribological contact (Fabricating Micromachines from Photoresist, April 2013). It was shown that a massive friction and wear reduction in such devices could be achieved by incorporating the hard-disk lubricants, perfluoropolyethers, into the SU-8 photoresist prior to use. The ability to simply fabricate SU-8 into microstructures suggests that this might provide a practical method of manufacturing cheap, sliding microcomponents.

Monitoring the Cutting Edge?

The lubricant film's thickness during machining is monitored by adding a fluorescent dye to the machining fluid

One of the central problems in understanding tribological phenomena is that they occur at a moving, buried, solid-solid interface. This means that it becomes exceedingly difficult to design *in situ* probes of the tribological interface. Thus, even apparently simple measurements of how much lubricant is present in the tribological contact, and how this varies with speed or load, are extremely difficult to make experimentally.

Such issues have been addressed for elastohydrodynamic lubrication by using ultra-thin-film interferometry, where a lubricant film is entrained between two relatively smooth surfaces. However, for rough surfaces under more severe conditions, this approach cannot be used. Thus, for example, it is still not clear how much lubricant, if any, can penetrate into the tool-chip contact during machining, making it difficult to answer questions concerning the quantity or type of lubricant that is required for adequate lubrication.

The groups of Srinivasan Chandrasekar and John Sullivan of Purdue University have addressed this question by measuring the fluorescence of a dye added to the lubricant [1]. Fluorescent materials absorb light at one frequency in the visible or ultraviolet region of the spectrum, but then emit light at a lower frequency. The difference in the frequency of the exciting and emitted radiation is generally quite large so that the incident radiation can be filtered out. The intensity of the emitted radiation is then proportional to the amount of fluorescent material (Fig. 1).

The experiment is carried out by adding 0.4 mg of a fluorescent material to 5 mL of a commercial cutting fluid and using this as a lubricant for a sapphire tool cutting a piece of 2 mm-wide lead. The interface between the stationary cutting tool and the moving lead block was uniformly illuminated using a UV-filtered mercury arc lamp and the sapphire "cutting tool" was shaped so that light could be extracted and imaged using a CCD camera.

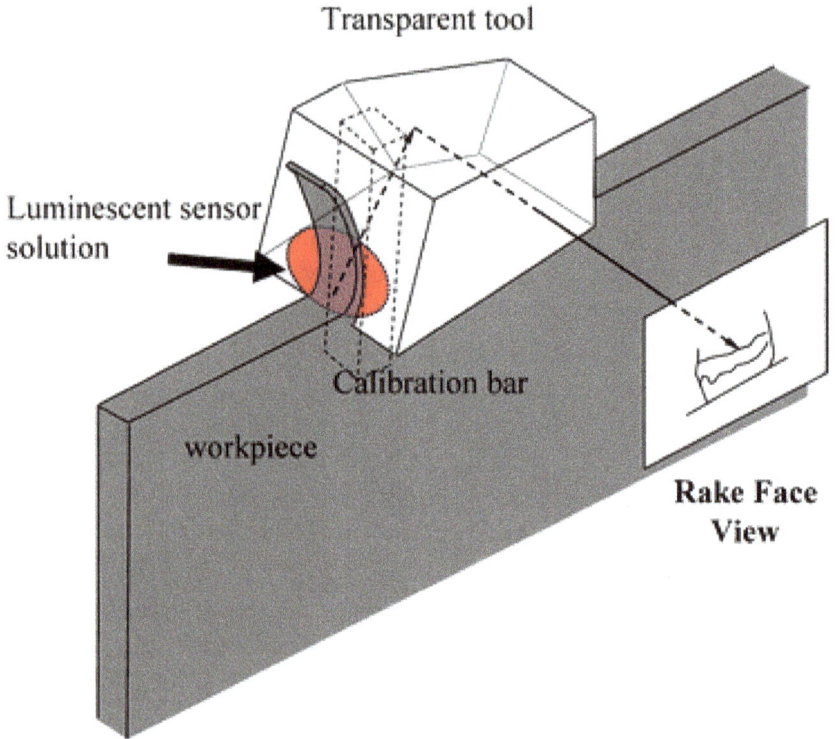

Fig. 1 Schematic of set-up for direct observation of tool rake face when machining with a sapphire tool. By kind permission of Springer Science+Business Media from Reference [1].

This enabled the variation in film thickness to be measured as a function of position by observing the variation of the fluorescence intensity in the image (Fig. 2). The intensity was calibrated as a function of film thickness by slightly tilting the cutting tool to produce a small gap where the thickness could be measured directly *in situ* during machining. This demonstrated that, indeed, the fluorescence intensity was proportional to thickness.

Carrying out similar experiments while sliding enabled the thickness of the lubricant film to be measured during machining, even though the interface was rough. It became apparent that the thickness of the film was very close to zero at the cutting edge, increasing almost linearly with distance, and reaching about 15 micrometers in thickness at 400 micrometers from the edge, corresponding approximately to the edge of the intimate contact zone.

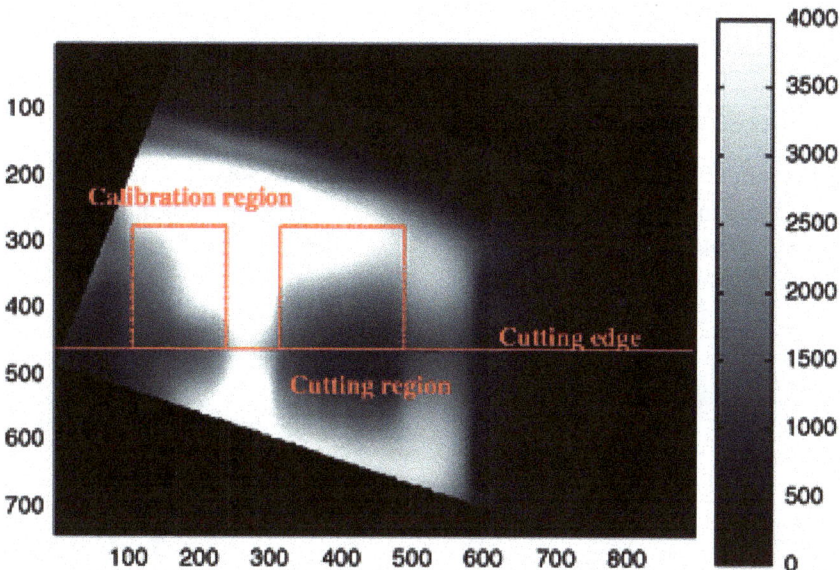

Fig. 2 Intensity image of rake face including calibration and cutting regions. The units of intensity are counts recorded by the CCD camera. By kind permission of Springer Science+Business Media from Reference [1].

While these preliminary measurements were made at relatively low sliding speeds, about 1 cm/s, this approach should also be applicable to machining at industrial cutting speeds. The authors also noted that this strategy could potentially be used to measure interfacial temperatures by using a phenomenon analogous to fluorescence, known as luminescence.

Luminescence is similar to fluorescence in the sense that incident high-frequency radiation is absorbed, and then emitted at a lower frequency. There are, however, subtle differences in the physics of luminescence, such that the time required to emit the radiation is much longer than for fluorescence — hence, for example, the persistent glow in luminescent watch dials.

More importantly, this decay time depends strongly on temperature and thus offers the possibility of simultaneously measuring both the lubricant film thickness, from the intensity, and the temperature, from the decay time, for a realistic tribological contact.

Tribology and Lubrication Technology
October 2007, 63(10) p64

Further Reading:

[1] Huang, C., Lee, S., Sullivan J.P. and Chandrasekar, S. (2007). In Situ Measurement of Fluid Film Thickness in Machining, Tribology Letters, 28, pp. 39–44.

Tribochemical Wear in a Transmission Electron Microscope

In-situ transmission electron microscopy studies reveal how chemical reactions affect the wear of diamond-like carbon

One of the central challenges in tribology is observing asperity contacts in a sliding interface. Electron microscopy can be used to image a contacting tip that mimics an asperity sliding against a surface, and a recent Cutting Edge article [1] showed how *in situ* electron microscopy could follow the fate of fullerene nanoparticles in the sliding contact.

The group of Professor Laurie Marks from Northwestern University, in collaboration with Dr. James Ciston from the Brookhaven National Laboratory and Drs. Ali Erdemir and Osman Eryilmaz from Argonne National Laboratory have extended this approach to examine the effects of the environment on the wear of diamond-like carbon (DLC) [2]. Films of diamond-like carbon grown by the Argonne group are some of the lowest-friction materials known, when measured in a vacuum or an inert environment, but are susceptible to wear when water vapor is present. Insights into the chemical effects that might contribute to wear were obtained by measuring electron-energy loss spectra (EELS) of the samples.

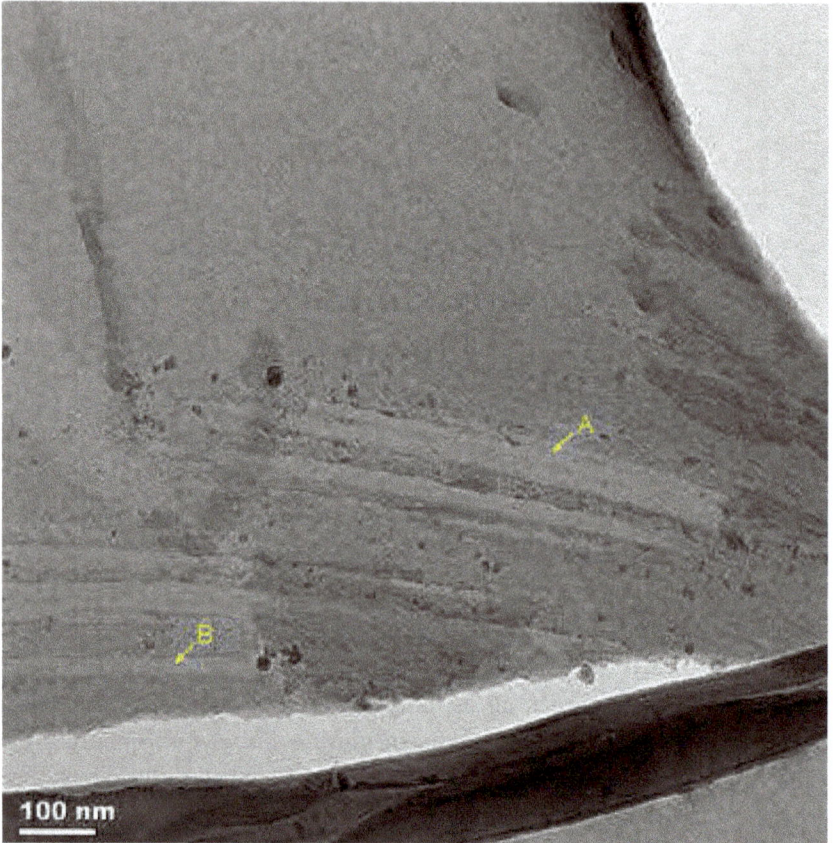

Fig. 1 An energy-filtered (zero-loss) micrograph showing the surface of the DLC film during a wet N_2 sliding experiment after 20 sliding passes. Arrows indicate lighter regions, which we interpret as being thinner due to the amorphous nature of the film. As A and B have different background brightness levels, separate calculations were done in each part of the micrograph in order to determine the estimated volume lost. By kind permission of Springer Science+Business Media from Reference [2].

In this technique, the energies of the electrons are analyzed after interacting with the sample. The characteristic energy losses of the electrons can be used to measure the ratio of sp^2 to sp^3 carbons in the sample, as well as to identify the elements that are present.

To prepare very thin, electron-transparent DLC samples, films were grown on a sodium chloride substrate. The water-soluble substrate allowed the films to be floated off and deposited onto a transmission electron microscope (TEM) sample holder. The DLC sample was rubbed against a tungsten tip with a radius of curvature of ~100 nm to mimic the asperity contact in a TEM that could operate with ~1.5 Torr of a gas present. Wear was followed by measuring changes in the contrast of the images, since thinner samples transmit more electrons and produce a brighter image.

Nanoscale wear tracks were found on the sample after rubbing in the presence of wet nitrogen (Fig. 1). The wear volume was found to increase with the number of sliding passes, but almost no wear debris was found. This suggested that wear had occurred by a tribochemical reaction to form volatile species.

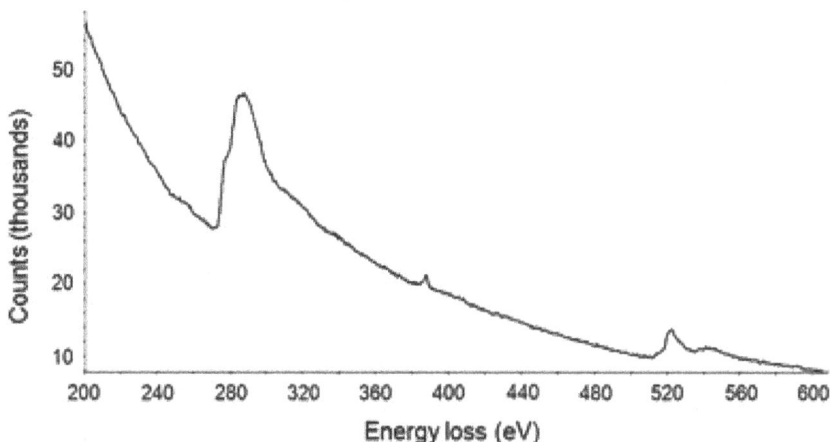

Fig. 2 EELS spectrum after 110 sliding passes in wet N_2 showing the relative prevalence of chemisorbed oxygen to nitrogen. From left to right, the carbon, nitrogen, and oxygen K-edges can be seen. Quantification suggests a two- to four-fold increase in the relative amount of oxygen compared to nitrogen present in the sampled volume over the course of a sliding experiment. By kind permission of Springer Science+Business Media from Reference [2].

The EELS spectra revealed both oxygen on the surface, which increased significantly as the samples were rubbed, as well as a growth of the proportion of sp^2-hybridized carbon, indicating that the surface becomes graphitized during rubbing (Fig. 2). In contrast, when the experiment was performed with wet hydrogen, no wear was found, although oxygen was still detected on the surface.

Evidently, the gas-phase environment has a significant influence on the wear of DLC. It was proposed that sliding produces chemically activated carbon atoms in the surface region. The activated atoms are passivated by hydrogen, but not by less reactive nitrogen. The activated carbon atoms are then oxidized by water from the gas phase to produce carbon monoxide, thereby leading to wear without debris formation. Such a combination of *in situ* imaging with chemical analyses of a sliding nano-asperity provides unique insights into the interaction between tribochemistry and wear.

Tribology and Lubrication Technology
February 2013, 69(2) p64

Further Reading:

[1] Tysoe, W. and Spencer, N. (2011). Imaging rolling nanoparticles, Tribology and Lubrication Technology, 67, pp. 96

[2] M'ndange-Pfupfu, A., Ciston, J., Erylmaz, O., Erdemir, A. and. Marks, L.D. (2013). Direct Observation of Tribochemically Assisted Wear on Diamond-Like Carbon Thin Films, Tribology Letters, 49, pp. 351–356.

Imaging Rolling Nanoparticles

*In-situ transmission electron microscopy studies reveal how
nanoparticles move in a sliding contact*

Ever since the discovery in 1985 of carbon "buckminsterfulleres", named for the inventor of geodesic domes that the structure resembles, the notion of using such molecules as nano-sized "ball bearings" has been appealing to tribologists. It has also been possible to synthesize inorganic analogs of these nanoparticle systems with other layered compounds, such as molybdenum or tungsten disulfides, and these have been shown to reduce friction and to be potentially useful oil additives or solid lubricants.

Since both MoS_2 and WS_2 are layered compounds, held together by weak van der Waals interactions, similar to those in graphite, the layers are able to easily slide over each other and provide low-friction films. However, the fact that the inorganic fullerenes reduce friction provides no insights into whether these operate as nano-sized ball bearings or whether they break up to form a thin lubricating layer on the surface.

While some information about what has occurred during sliding can be obtained using post-mortem analyses, questions regarding what happens at a sliding-sliding solid interface can only really be answered by interrogating the interface during sliding. Unfortunately, techniques that can probe such interfaces are few and far between.

To address this question, Dr Fabrice Dassenoy, working in the group of Professor Jean-Michel Martin at the École Centrale in Lyon, France, in collaboration with the "Nanofactory" company, have designed a novel

147

sample holder that can be incorporated into a transmission electron microscope (TEM), which allows them to look directly at inorganic fullerene nanoparticles in a sliding contact [1] (Fig. 1).

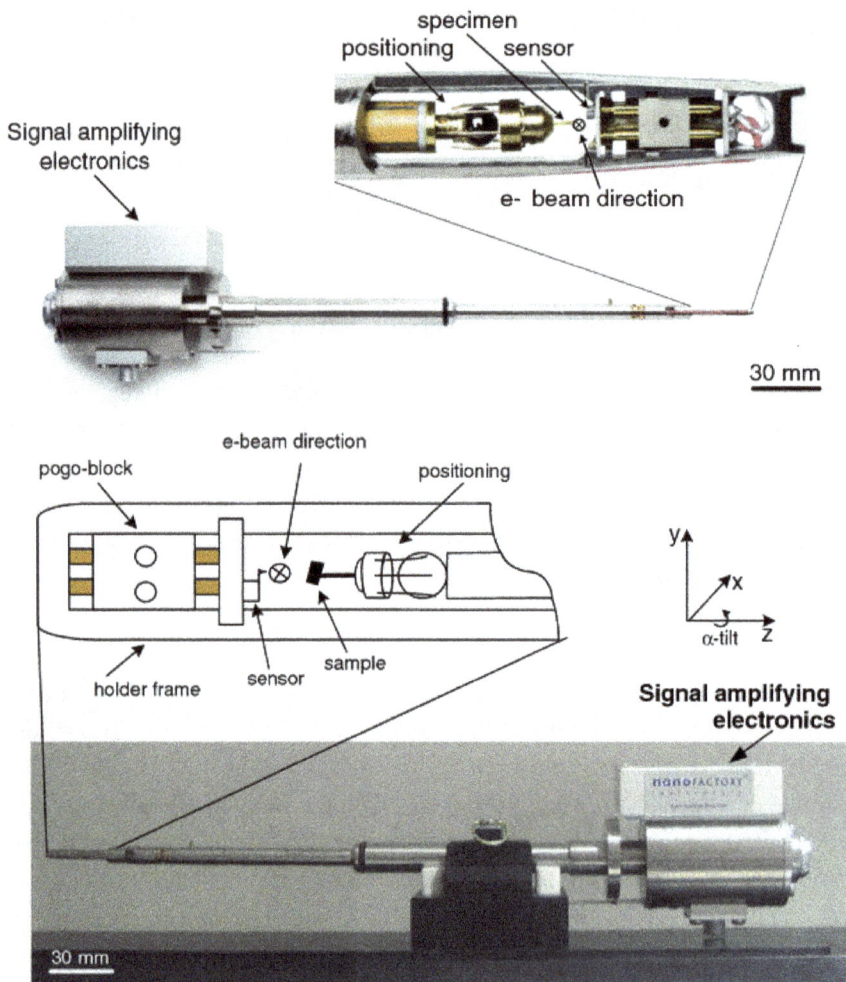

Fig. 1 Schematic of the TEM–AFM holder design. By kind permission of Springer Science+Business Media from Reference [1].

In order to make an electron-transparent substrate, required for TEM, they used a silicon wedge. Sliding was achieved using a flattened atomic force microscope (AFM) tip, with a flat area of about 500 nm^2, and different normal forces could be applied while sliding.

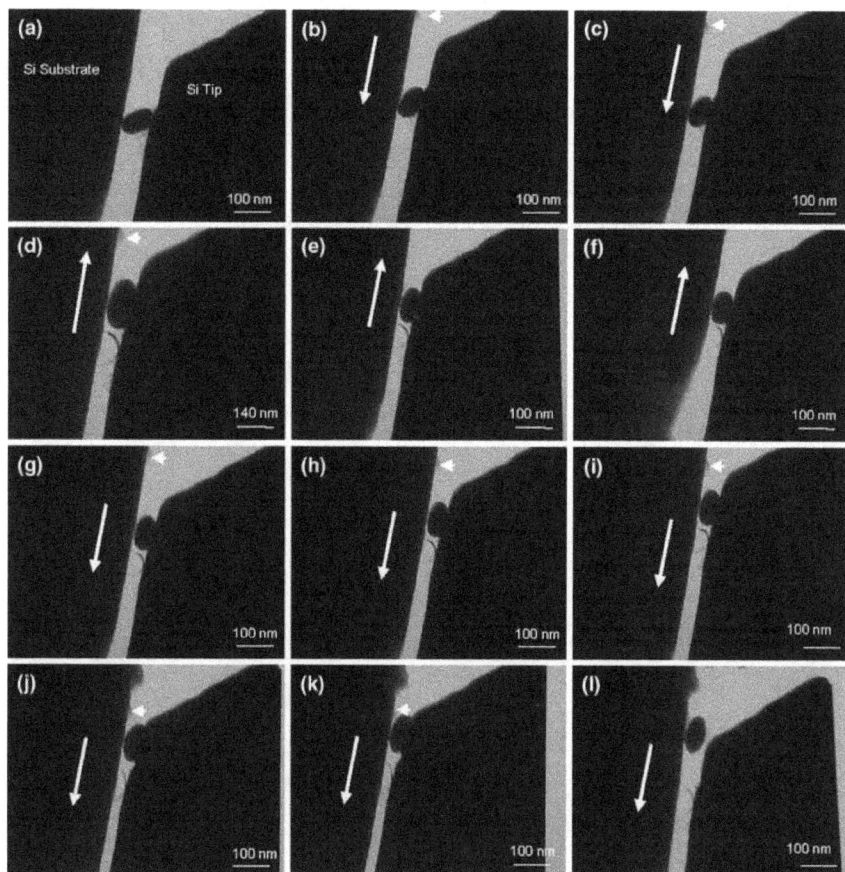

Fig. 2 TEM images showing a particle of IF-MoS$_2$ and a piece of the particle resulting from exfoliation of the particle. By kind permission of Springer Science+Business Media from Reference [1].

Experiments were carried out using highly crystalline molybdenum disulfide nanoparticles of about 70 nm in diameter that had been synthesized by Professor Reshef Tenne's group, from the Weizmann Institute in Israel — pioneers in the use of inorganic fullerene nanoparticles. They first examined what happened to a nanoparticle under a relatively low normal force of 100 nN. In this case, they were able to follow a nanoparticle rolling about half of its circumference, both in the forward and backward directions, without any apparent damage to the particles.

They then examined the effect of increasing the normal force to 400 nN with a slightly larger particle diameter of 100 nm. Now they found that the particle stuck to the tip without sliding. In addition, they were able to discern the appearance of some layered material, suggesting that the nanoparticle had undergone some exfoliation (Fig. 2).

While the authors acknowledge that these are preliminary results that require much more systematic investigation, the results do suggest that the nanoparticles can operate both by rolling and by exfoliation. More importantly, however, they show that we now have to the tools to peer directly into the sliding solid-sold interface and see tribological effects as they occur.

Tribology and Lubrication Technology
June 2011, 67(6) p96

Further Reading:

[1] Lahouij, I., Dassenoy, F., de Knoop, L., Martin, J.-M. and Vacher, B. (2010). In Situ TEM Observations of the Behavior of Individual Fullerene-Like MoS_2 Nanoparticles in a Dynamic Contact, Tribology Letters, 42, pp. 133–140.

The Gold Standard for Wear

A Canadian team of researchers has developed a method of measuring component wear that doesn't involve radioactive materials

The measurement and minimization of wear are often central in designing mechanical components. Unfortunately, the very low wear rates that are the most interesting and useful are also those that are the most difficult to measure precisely.

For example, modern, high-efficiency engines have wear rates that are on the order of nanometers/hour. One of the most accurate methods is based on radionuclide techniques, which involve using an accelerator to generate a surface region that is radioactive and then measuring the amount of radiation in the lubricant, caused by the wear process. This approach is useful for continually measuring wear rates *in situ*, but there are obviously many environments where using radioactive materials is inconvenient.

A team led by Professor Peter Norton at the University of Western Ontario, Canada has recently developed an analogous approach that avoids using radioactivity [1,2]. This method involves implanting a surface region with gold atoms, (^{197}Au) at energies chosen to match the implantation depth to the desired measurement parameters. For example at 90 keV, gold ions penetrate a few tens of nanometers into ferrous metals and alloys. Rather than using radioactivity to monitor material removal as is done with radionuclide techniques, they use Rutherford backscattering (RBS) (Fig. 1).

RBS was originally developed in the early part of the 20th century by Rutherford who used it to disprove the "plum-pudding" model of the atom, which postulated that the atom consisted of negative electrons immersed in a sea of positive charge. It was found that positively charged α-particles incident on a gold foil were scattered in a backward direction, which was only possible if all the positive charge of the atom was concentrated in a small positive "nucleus", thereby invalidating the "plum-pudding" model and paving the way for understanding atomic and molecular structure.

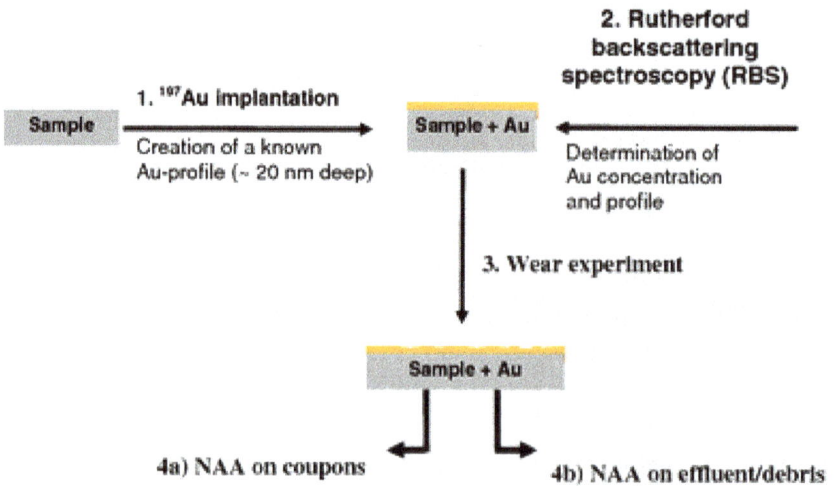

Fig. 1 Strategy of the low wear rate method combined Au implantation, RBS, and NAA techniques.

1) The implantation of [197]Au with a shallow, known low-concentration profile into 52100 steel or 1095 carbon steel.

2) Determination of the [197]Au profile and concentration using carbon Rutherford backscattering spectroscopy (C-RBS).

3) Performance of the wear tests.

4) Determination of the [197]Au profile and concentration by C-RBS after the wear tests and/or collection of debris and determination of Au concentration using neutron activation analysis (NAA).

By kind permission of Springer Science+Business Media from Reference [1].

The energy of the back-scattered particles from a given depth in a given material depends on the mass of the nucleus. The Ontario group used this approach, rather than radioactivity, to measure material removal from the film by back-scattering carbon ions and measuring their energy to distinguish gold from other elements, for example, iron, copper, silicon etc., in the sample.

The ions also lose energy as they pass through the sample, so that careful measurements of the energy distribution of the backscattered ions from the gold can be converted into a depth profile of the gold that has been implanted into the sample. This in turn can be converted into a calibration of the amount of gold lost from the sample *versus* the worn depth. Because of the small gold implantation depth, this curve is very sensitive to wear of the top few tens of nanometers and is thus ideal for ultralow wear measurements even in the presence of surface films.

Since ion implantation is well known to modify materials' mechanical properties, the gold concentration was limited to ~0.1 to 1% of that typically used for surface modification, and measurements of mechanical properties using an indenter showed no statistical differences between implanted and non-implanted samples.

Wear experiments were carried out in a pin-on-flat configuration using AISI 52100 steel with a mineral oil alone, and containing a typical engine anti-wear additive, zinc dialkyldithiophosphate (ZDDP). No statistical differences in friction coefficient were found for samples containing various doses of implanted gold.

However, the samples were rapidly depleted of gold when rubbing in just the base oil showing an initial wear rate of ~1.7×10^{-7} mm^3/N m after about 1 minute, quickly decreasing to about 1×10^{-8} mm^3/N m after 30 minutes of sliding. Similar experiments were carried out for an oil with 1.2% ZDDP. The amount of gold removed from the sample was significantly reduced, from about 27% when using only the base oil, to less than 5% when ZDDP was added. The ability to measure the depth profile of the gold also enabled changes due to plastic deformation or film growth to be monitored at the same time as measuring low wear rates.

Tribology and Lubrication Technology
February 2009, 65(2) p88

Further Reading:

[1] Li, Y.-R., Shakhvorostov, D., Pereira, G., Lachenwitzer, A., Lennard, W.N. and Norton, P.R. (2009). A Novel Method for Quantitative Determination of Ultra-low Wear Rates of Materials, Part I: On Steels, Tribology Letters, 33, pp. 143–152.

[2] Li, Y.-R., Shakhvorostov, Lennard, W.N. and Norton, P.R. (2009). A Novel Method for Quantitative Determination of Ultra-low Wear Rates of Materials, Part II: Effects of Surface Roughness and Roughness Orientation on Wear, Tribology Letters, 33, pp. 63–72.

Stuck in a Rut

Structuring surfaces on a mesoscopic scale leads to precise friction control

Over the last decade, micro- and nanoelectromechanical systems (MEMS and NEMS) have become increasingly commonplace in a wide variety of devices with applications from automobile air-bag accelerometers to their use in a host of consumer electronic products. As we have reported before [1], tribological problems in such devices have been an issue, and controlling friction and wear still remain a challenge. While our previous column on this subject dealt largely with wear, a recent paper [2] in the STLE-affiliated journal *Tribology Letters* deals with the control of friction on length and load scales relevant to microdevices.

At the macroscopic scale, many researchers have explored the effects of surface structure on lubrication. Structures of the appropriate dimensions have been shown to improve hydrodynamic lubrication while also serving as a sink for wear particles, thus rendering them less likely to cause wear by plowing. Structuring at the micro and nano scales has been explored far less, and tribological measurements involving contact sizes and loads comparable to those present in MEMS and NEMS devices (at the "mesoscopic scale") have been rare. While atomic force microscopy (AFM) can achieve the loads that are relevant to MEMS, it generally involves sharp tips with contact areas on the nanometer-scale.

To address the contact-area problem, Johannes Sondhauß and Harald Fuchs from the University of Münster, Germany, together with André Schirmeisen from the University of Giessen, Germany have explored the effect of structuring on friction by using so-called "colloidal probe" AFM [3], where the usual sharp tip is replaced by a sphere with a diameter of micrometers.

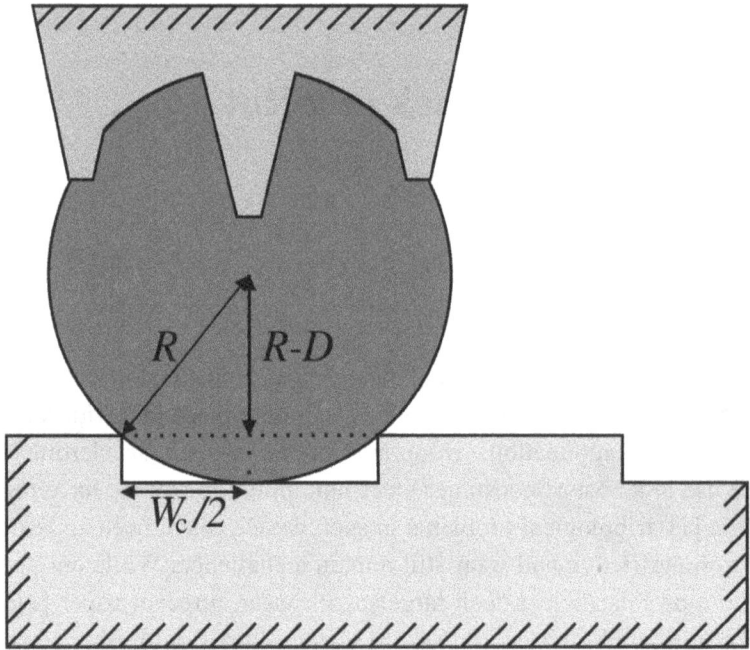

Fig. 1 Geometrical overview of the artificial tip-sample contact. By kind permission of Springer Science+Business Media from Reference [2].

The technology that was crucial to this study was the focused-ion beam (FIB) technique, which allows the precise milling of materials with a resolution of nanometers. The authors used FIB to prepare a library of flat-bottomed groove arrays on a single silicon wafer. The groove took up exactly half of the wavelength of a square-wave-like cross section, and the groove width was varied from 530 nm to 4350 nm in 23 steps. The depth of the groove in all cases was 26 nm. The tip consisted of a titanium sphere (either 4.6 or 15.8 μm in diameter) that was glued into a hollowed-out AFM tip (FIB again!). Friction measurements were carried out in air, with the sliding direction perpendicular to the groove axis. The effective friction coefficient (calculated as the slope of the linear friction force-applied load curve) was averaged over an area of hundreds of square microns (i.e. across several grooves at many positions).

The results showed that, for both sizes of tips, tested against the whole range of groove spacings, there existed a maximum friction at a groove width corresponding exactly to the width at which the sphere interlocks with the groove, simultaneously touching both walls and the base. This can be calculated by simple geometry. Focusing in on the friction as the sphere traverses a particular groove, it was observed that for a groove width where interlocking occurred, a higher friction coefficient was instantaneously observed than on the neighboring terrace.

FIB is a valuable addition to our arsenal of techniques for tribological experimentation. It allows the precise sculpting of structures on the nanometer scale, and therefore enables us not only to mimic the roughness of engineering surfaces, but allows us to craft new tools for the investigation of contact in a new generation of mechanical devices.

The results showed that unidirectional sliding (i.e. no direction reversal of the wheel) led to up to 40% higher wear than cases where the rotation direction was reversed. Beyond five cycles of changing direction, the effect did not increase significantly. Overall, the Vickers' hardness decreased as the number of direction-change cycles was increased. XRD measurements of unidirectionally *versus* bidirectionally (50 cycles) worn surfaces showed negligible differences, suggesting that textural changes in the alloy did not play a major role in the wear differences. On the other hand, subsurface cross-sections measured by SEM showed that the unidirectionally worn samples had suffered more severe fracture than the bidirectionally worn materials. Fracture involves the nucleation and propagation of cracks, which occur when stress concentrations exceed the critical stress at fracture. The stress concentrations arise when dislocations of the same sign interact with each other, piling up at interfacial boundaries. Due to the Bauschinger Effect, local reversible movement of dislocations may occur and the number of dislocations is reduced in bidirectional sliding, as dislocations become annihilated by dislocations of opposite sign, which arise from sliding in the opposite direction. Furthermore, bidirectional sliding may not only reduce crack formation, but it may also slow crack propagation, since the stress driving the crack is also reversed.

There are interesting consequences of these observations. Firstly, design of machines that minimize unidirectional in favor of bidirectional sliding could lead to reduced wear. Secondly, abrasive processes used for manufacturing could be improved if unidirectional sliding were favored over bidirectional sliding.

Tribology and Lubrication Technology
August 2011, 67(8) 72

Further Reading:

[1] Tysoe, W. and Spencer, N.D. (2008). Alcohol Gets You no Wear, Tribology and Lubrication Technology, 64, p64.

[2] Sondhauß, J., Fuchs, H. and Schirmeisen, A. (2011). Frictional Properties of a Mesoscopic Contact with Engineered Surface Roughness, Tribology Letters, 42, pp. 319–324.

[3] Ducker, W.A., Senden, T.J. and Pashley, R.M. (1999). Direct Measurement of Colloidal Forces using an Atomic Force Microscope, Nature, 353, pp. 239–241.

Fabricating Micromachines from Photoresist

Simple additions of a common hard-disk lubricant and a nanoparticle filler to photoresist open new tribological applications for microfabricated devices.

We have previously reported on microelectromechanical devices (MEMS) and the trials and tribulations involved in reducing friction and wear in sliding micromachines constructed from silicon [1]. Because of its use in electronic devices, silicon is, of course a highly versatile material for microfabrication — by means of photolithography and etching processes. However, there is increasing interest in using a photoresist material, known as SU-8, itself to fabricate micromachines. SU-8 is a negative, thick-film uv photoresist based on epoxy resin, a solvent, and a photoacid generator, which is what imparts the photosensitivity. Fabricating microstructures from SU-8 is much simpler than using silicon, since no etching is necessary; making MEMS devices requires SU-8 to be spin-coated on a substrate, exposed to uv light through a mask to cross-link it, and the non-crosslinked photoresist to be removed, followed by some heat-treatment steps.

There is, of course, a catch: while SU-8 is convenient to use, it has poor mechanical and tribological properties — low modulus, low hardness, high friction and poor wear resistance. However, in a recent paper in STLE-affiliated journal *Tribology Letters*, Prabakaran Saravanan, Nalam Satyanarayana, and Sujeet K. Sinha, from the National University of Singapore, have described a novel approach to

159

improving these properties significantly [2]. The approach was two-pronged: a well-known lubricant used by the hard-disk industry, hydroxyl-terminated perfluoropolyether (PFPE), was mixed with the SU-8 prior to spin-coating. Thanks to the presence of hydroxyl groups in the PFPE, these lubricant molecules reacted with the epoxy groups of the SU-8 during curing, becoming incorporated throughout the crosslinked structure. To further boost the mechanical properties, nanoparticles are also mixed into the SU-8.

Fig. 1 (Digital Image of a 200 μm thick gear made of SU-8 + PFPE composite, using a UV lithographic process.) By kind permission of Springer Science+Business Media from Reference [2].

The effects of PFPE incorporation on the tribological properties of the SU-8 were dramatic: the friction coefficient of SU-8 measured against Si_3N_4 in air was a hefty 0.82, while SU-8 with PFPE incorporation displayed a value of 0.09! The wear resistance of the material was also increased by a factor of 10^4.

Experiments with nanoparticles alone (silica, carbon nanotubes, or graphite) incorporated into the SU-8 showed negligible improvement of nanotribological properties. Moreover, the nanoparticles showed no improvement in modulus and actually reduced the hardness, compared to pure SU-8. Interestingly, a combination of PFPE and nanoparticles showed similar improvement in friction and wear behavior to the SU-8 + PFPE composites, with enhancements in modulus ($\times 1.4$) and hardness ($\times 1.4$).

The authors explain these interesting results in terms of the distribution of PFPE throughout the matrix, rendering it available to the sliding contact, even if wear should occur. The presence of PFPE on the surface, even after 10^6 sliding cycles, was confirmed by both x-ray photoelectron spectroscopy and water-contact-angle measurements.

The ease with which this tribologically enhanced SU-8 can be fabricated suggests that it might have a bright future in applications such as bearings, raceways, gears, biodevices, precision-positioning stages, and components in consumer electronics, such as cameras and printers.

Tribology and Lubrication Technology
April 2013, 69(4) p64

Further Reading:

[1] Tysoe, W. and Spencer, N.D. (2008). Alcohol Gets You no Wear, Tribology and Lubrication Technology, 64(4), p. 64.

[2] Saravanan, P., Satyanarayana, N. and Sinha, S.K. (2013). Self-lubricating SU-8 Nanocomposites for Microelectromechanical Systems Applications, Tribology Letters, 49, pp. 169–178.

Topic 6

Biotribology

Looking back over the last decade, it is remarkable how many columns we wrote on biological aspects of tribology. While on the one hand it is clear that lubrication of artificial joints is an important medical issue in an aging population, this was by far not the only topic that caught our imaginations. In fact, the first biotribological piece we wrote was on teeth (*Tribological Testing of Teeth*, June, 2006), reporting on the potential of tungsten disulphide nanoparticles, incorporated into nickel-phosphorus electroless films, to lower the friction between braces and wires, thereby facilitating orthodontic procedures.

Epizoochory is the mechanism by which plant burrs stick to animal fur. Natural mechanisms can serve as interesting models for new man-made adhesion systems, and epizoochory was the inspiration for Velcro®, which was the subject of a mechanics study in Tribology Letters and inspiration for a TLT column (*Sticking with Epizoochory*, April, 2007). A more recent example was also covered in the same column, namely the extraordinary ability of geckos to stick upside down on ceilings. Studies of these animals have led to efforts to mimic the impressive nanoscale structures on geckos' feet to make new artificial adhesive systems, purely based on physical forces.

Important though tribology is in many areas of medicine, it is often extremely hard to determine friction in biological situations. One example is during the implantation of stents in blood vessels, where the frictional force is significant but hard to predict or measure. Another is the friction between a contact lens and the eyelid, which is widely believed to correlate with lens comfort. Attempts have been made to measure and to model such systems. In *Fenders and Stents*, August 2007, we described measurements of friction on a layer of aortic endothelial cells under serum. Not only could a friction coefficient μ be measured in this setup, but the onset of cell damage could also be correlated to the applied normal load. In a subsequent column (*Sliding on Cells*, August, 2010), we discussed the mechanical model necessary to understand such systems — the elastic foundation model — which enables the load-dependent shear stress at the cell surface to be determined. A slightly different approach was described in *Keeping up with Contacts*, April, 2012, where a method for the measurement of contact lens friction was described that relied on the surface functionalization of polymeric counter surfaces with natural molecules such as mucin. In this way, the surface of the eyelid could be mimicked. Such measurements allowed the quantitative investigation of contact-lens properties, and the tribological behavior of different lens technologies to be compared.

Hip implants are still a major area of scientific investigation, but it is surprising how inconsistent the literature is in the entire area of hip lubrication — both natural and artificial. In *Hip Surface Science*, August, 2009, we highlighted a review article in Tribology Letters that set out to correct a number of misconceptions that might have been gleaned from the literature. A typical example is that many papers have reported the structures of cartilage surfaces that are entirely due to artifacts introduced by sample preparation or vacuum exposure during investigation in an electron microscope! The cartilage-cartilage interaction was mimicked with hydrogel sliding partners in *Sliding Soft Surfaces,* June, 2014, in which the differences between gel-gel friction and that measured between a gel and an impermeable counter-surface were described. The highly hydrated nature of the gels led to an almost completely speed-independent friction behavior when they were slid against each other, while the friction decreased significantly with speed, when a glass ball was slid on a flat gel surface, for example.

Finally, in *How Many Licks? The Lollipop and Biotribocorrosion*, December, 2014, a careful study involving lollipop lickers, backed up with detailed lollipop metrology and Monte Carlo modeling approaches, was used to mimic biotribocorrosion — a highly important phenomenon in both implantology and marine engineering. Among other revelations: lickers appear to adopt only one of two completely distinct licking styles!

Tribological Testing of Teeth

Metal chalgogenide nanoparticles are found to effectively lubricate dental braces

Approximately four million people in the United States wear braces to straighten their teeth. Orthodontics aims to change the location of abnormally aligned teeth by applying a constant force of between 1 and 2 Newtons. This force leads to a continuous degradation and regeneration of the tissue supporting the teeth (known as the "periodontal ligament") so that the teeth are moved to new, and presumably more aesthetically pleasing, positions.

This force is achieved by placing a rigid wire through slots in the braces and then attaching elastic or springs to the braces as well (Fig. 1). Frictional forces are present between the wire and the slots as the teeth are being straightened.

Moreover, as the teeth move, they have a tendency to tilt so that at some critical angle, the wire makes contact with the edge of the slot, causing adhesion and an increase in friction. At even higher angles, plastic deformation can occur and may even completely stop the tooth from sliding.

Significantly higher forces must be applied to the teeth to overcome friction, with several consequences. First, the force on each tooth eventually becomes uneven as each tooth tilts at a different rate, causing a lengthening of the treatment and frequent dental visits to "fine tune" the orthodontic appliances.

Second, to apply a force to move the teeth, other teeth must be anchored. This is generally achieved by applying the force to the back teeth (the molars), but excessive force can also cause these to move. Such unwanted motion can be avoided by anchoring to a number of teeth, or by drilling mini-implants directly into the bone and applying the force to the implants.

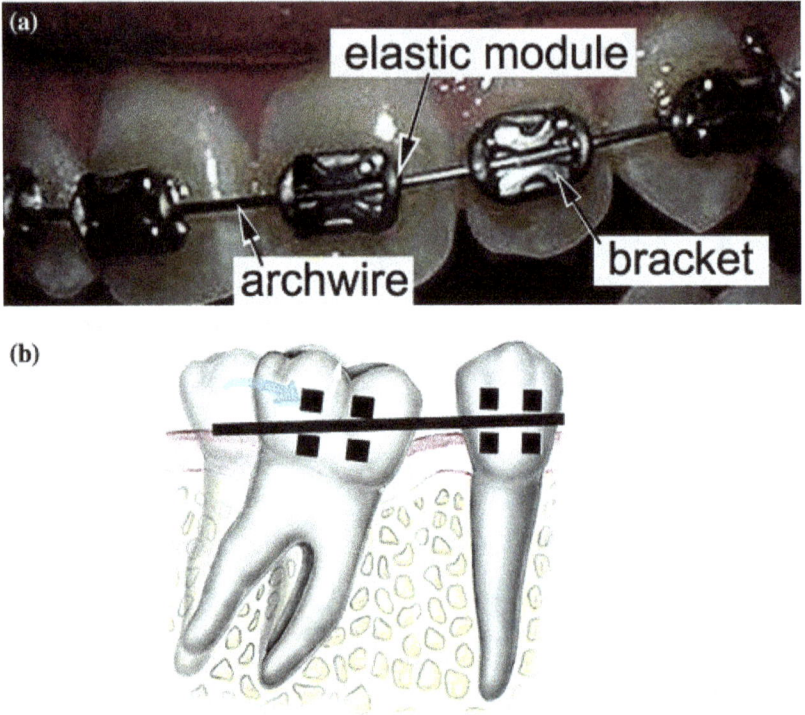

Fig. 1 (a) Photograph showing the orthodontic setup of a fixed appliance; (b) schematic illustration of the tilting movement of a tooth, which brings about a growing pressure between the archwire and the corners of the bracket slot.By kind permission of Springer Science+Business Media from Reference [1].

Numerous attempts have been made to ameliorate these problems by redesigning the braces. The group of Reshef Tenne at the Weizmann Institute in Israel, in collaboration with Drs. Redlich and Katz from the Department of Orthodontics, Hebrew University-Hadassah Medical School, however, decided to address the tribological problem directly [1]. This group has pioneered the field of inorganic fullerenes and subsequently together with Lev Rapoport, Holon Institute of Technology, their use as solid lubricants.

Fullerenes have a molecular structure that looks like soccer balls made up of pentagons and hexagons bound together into a round, hollow molecular cage, resembling the geodesic domes created by the architect R. Buckminster Fuller. Robert Curl and Richard Smalley of Rice University first identified this unique structure in carbon in 1985 and received the 1996 Nobel Prize in Chemistry for their discovery.

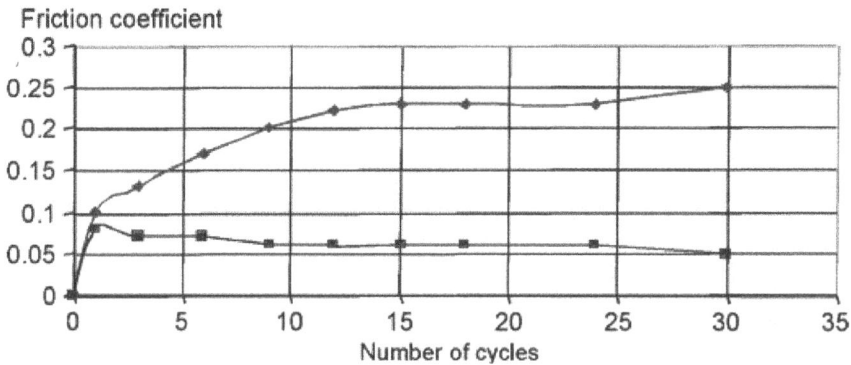

Fig. 2 Tribological tests were performed using a ball-on-flat device with a sliding velocity of 0.2 mm/s and a load of 50 g (ca. 1.5 GPa). A ball bearing with a diameter of 2 mm was used as a counter body. Dry and wet friction tests with paraffin-oil lubricant were carried out during 50–200 cycles. The hardness of the Ni–P coating was close to 7500 MPa. This relatively high hardness usually led to formation of ploughing tracks on the surface of the ball. Blue curve is the uncoated stainless steel wire; red curve is the wire coated with Ni–P and IF-WS$_2$ nanoparticles. By kind permission of Springer+Business Media from Reference [1].

The metal chalcogenide analogs of these carbon compounds were discovered by Tenne's group in the early 90's and were found, under certain conditions, to serve as solid lubricants. The Tenne group theorized that these might also acts as a solid lubricant in dental applications, if only they could be released slowly. The apparent absence of deleterious biological effects of inorganic fullerenes also makes them promising candidates; preliminary experiments with rats show that they are non-toxic and lead to no ill effects upon skin contact.

The release problem was solved by incorporating tungsten disulfide fullerene nanoparticles of about 120 nanometers in diameter into a nickel-phosphorus electroless film. These films have excellent wear and corrosion resistance and can be conveniently plated onto metals from solution.

Tribological tests were carried out between braces and wires oriented at various angles to mimic the tilting referred to above (Fig. 2). While the friction reduction when the slots were not tilted was modest (about 17 %), in the most important range, at higher tilt angles, friction was less than half that of the uncoated wires. At low angles, the nanoparticles are suggested to act as spacers that reduce the number of asperities that come into contact, thereby lowering friction. At higher angles, the nanoparticles are released from the nickel-phosphorus film to form lubricious tungsten sulfide nanosheets that lead to facile sliding. It appears that tribology can lead not only to improvements in medical procedures, but also to fewer visits to the dentist!

Tribology and Lubrication Technology
June 2006, 62(6) p88

Further Reading:

[1] Katz, A., Redlich, M., Rapoport, L., Wagner, H.D. and Tenne, R. (2006). Self-lubricating Coatings Containing Fullerene-like WS_2 Nanoparticles for Orthodontic Wires and Other Possible Medical Applications, Tribology Letters, 21, pp. 135–139.

Sticking with Epizoochory

Nature has developed sophisticated adhesion mechanisms that have provided inspiration for man-made devices

From time to time, we like to publish papers on adhesion in Tribology Letters. Adhesion is obviously a close cousin to tribology, since many important issues, such as contact phenomena, are highly relevant to both fields. A recent such paper from the group of John Williams at Cambridge University, UK, deals with probabilistic fasteners [1]. These are probably better known as Velcro®, which is a 50-year-old biomimetic invention involving interlocking hooks and loops. The design is based upon a natural phenomenon known as *epizoochory* — the mechanism by which plant burrs attach to animal fur. This attachment mechanism is probabilistic, in the sense that one side of the contact consists of randomly oriented elements, so that the two sides do not necessarily need to be precisely aligned to allow adhesion. The paper from the Williams group describes mechanical measurements on Velcro®, and concludes that approximately 50% of the hooks contribute actively to resisting the load applied to separate a joint.

Many adhesive contacts are probabilistic, in the sense that precise alignment between the surfaces is not necessary. In the cases with which we are most familiar (e.g. adhesive tape), the adhesive elements are on the molecular scale. In such cases, of course, the mechanical properties of the backing tape and the viscoelastic properties of the glue layer are key to the tape's ability to conform and stick to surfaces, with the largest possible contact area. Again using nature as a guide to improving adhesion, it has also been found that the impressive adhesive properties of a gecko, which are strong enough to allow it to hang upside down on a ceiling, are due to nano-sized, adhesive hairs or *spatulae* on the gecko's foot. These are held within a conforming structure that maximizes the number of *spatulae* contacting the surface on which the gecko is walking. A group headed by Eduard Arzt from the Max-Planck-Institute (MPI) for Metals Research in Stuttgart, Germany have used the knowledge gained from the study of the adhesion of geckos and smaller animals to establish design rules for better adhesives [2,3].

The MPI researchers showed that the smaller the contacting fibers, the larger the adhesive force (due to van der Waals interactions) that can be obtained per unit area — a phenomenon known as "contact splitting". For relatively heavy animals such as the gecko, large adhesive forces are critical, of course. Correspondingly, it was found that there seems to be an inverse correlation between the size of the animal and size of the adhesive structure necessary to support its weight (Fig. 1). While this is a good start to the design process, it has to be remembered that there is a lower limit to the size of the fibers, below which they fracture, rather than detaching from the surface reversibly. Another problem with reducing fiber size is that inter-fiber adhesion can occur if they are not sufficiently stiff, reducing the effectiveness of contact splitting. The Arzt group proposes constructing "adhesion maps", by logarithmically plotting, over many orders of magnitude, the fiber radius, r, against the Young's modulus, E. Plotting lines on the map that describe the phenomenon of inter-fiber adhesion, the ability to conform to the surface, and the apparent contact strength of the fibers as they pull off the surface (itself a function of Young's modulus), the MPI group identified a triangular region within the r vs E plot that corresponds to fibers that maximize the contact, and therefore adhesion. Biological contact elements appear to lie within this triangle, which lends credence to the approach.

An interesting consequence of this is that low densities of adhering fibers are conducive to higher contact strength, since they help circumvent the inter-fiber adhesion. The next step in the process is the use of such fiber-based adhesion design maps to create new adhesion systems.

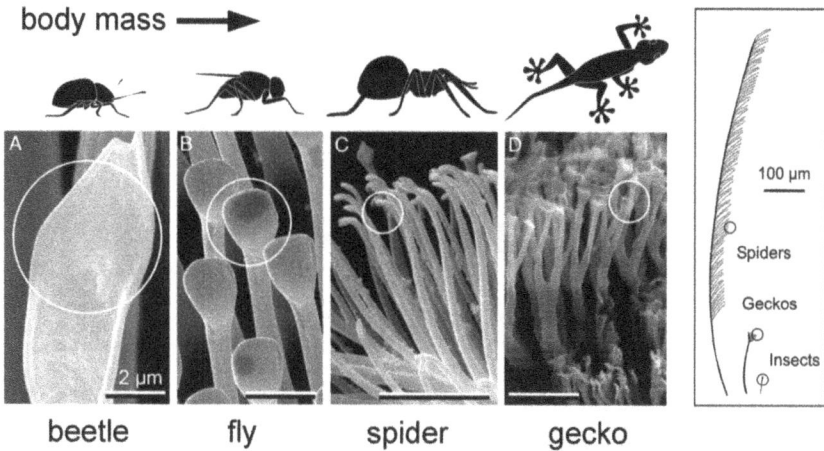

Fig. 1 Terminal adhesive elements (circles) in animals with hairy design of attachment pads. By kind permission of Proceedings of the National Academy of Sciences, from Reference [2]. Copyright (2003) National Academy of Sciences, U.S.A.

As we have seen before in tribology, this is an example where the combination of different disciplines, namely physics (E. Arzt), biology (S. Gorb) and materials science (E. Arzt and R. Spolenak (now ETH Zurich)), results in significant advances in our understanding of mechanisms and our ability to develop new technologies.

Tribology and Lubrication Technology
April 2007, 63(4) p56

Further Reading:

[1] Williams, J.A., Davies, S.G. and Frazer, S. (2007). The Peeling of Flexible Probabilistic Fasteners, Tribology Letters, 26, pp. 213–222.

[2] Arzt, E., Gorb, S. and Spolenak, R. (2003). From micro to nano contacts in biological attachment devices, Proc. Nat. Acad. Sci., 100, pp. 10603–10606.

[3] Spolenak, R., Gorb, S. and Arzt, E. (2005). Adhesion design maps for bio-inspired attachment systems, Acta Biomaterialia, 1, pp. 5–13.

Fenders and Stents

Arterial stents are life-giving devices implanted into millions of patients, and the friction between the stent and the wall of the artery is a crucial parameter in their performance. However, it has been hardly investigated until now.

As we all know, tribology is everywhere, and it is hard to think of a technology where it isn't involved at some level. Nevertheless, even though friction is often critical to the performance of a manufacturing process, that doesn't mean the tribological behavior of the process is understood. Often, nowadays, complex processes such as forming of automobile fenders are modeled by a sophisticated finite-element approach. However, the values of friction coefficient that are included in such calculations may have to be virtually plucked out of thin air, especially when lubricated friction is involved, simply because they have never been measured and reliable mathematical models of boundary lubrication are not available.

An extreme case of guesswork in modeling has been in the area of arterial stents. Many owe their lives to the widespread surgical procedure known as coronary angioplasty, during which a balloon is inflated within an artery, in order to prevent it from blocking. A metal structure, known as a stent, is then expanded within the region, in order to hold it open after the balloon has been removed. Much modeling effort has been put into understanding stenting systems, since the mechanical properties and hence the design of the stent, as well as the application pressure, need to be optimized. Too little pressure may lead to subsequent displacement of the device due to blood flow, with potentially disastrous consequences. Too much pressure could damage the epithelial cells in the arterial wall. One well-known modeler of stents has actually admitted that "…we carried out the simulations without friction, since reliable coefficients describing the frictional behavior are not available yet".

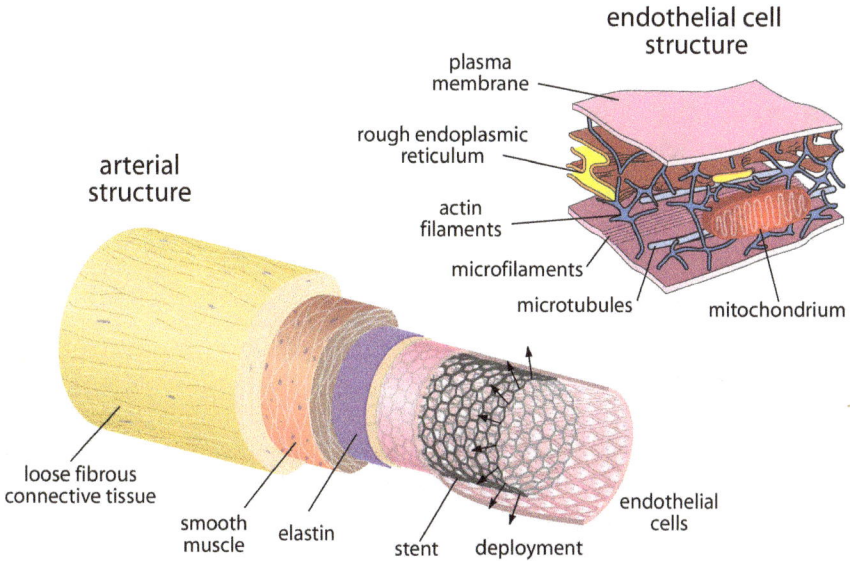

Fig. 1 (bottom left) Schematic of a stent located inside of the arterial structure. (Upper right) Schematic of the structure of an endothelial cell including the structural elements such as the actin filaments (blue), microfilaments (light blue), and microtubules (lining the bottom plasma membrane) By kind permission of Springer Science+Business Media from Reference [1].

Enter Professor Greg Sawyer, of the University of Florida, and his team. In a recent issue of Tribology Letters they reported a novel approach for measurement of friction against cell layers, generating, for the first time, the important μ values of inorganic materials slid against a layer of aortic epithelial cells [1]. The tribometer and the procedure that they used were anything but standard: the load range was at the mN level and below, the stroke length was 500 μm, the measurement was carried out in a Petri dish under serum, and the cells were stained and analyzed by fluorescence microscopy after tribological testing.

The results, obtained at normal loads of around 0.4 mN show that μ ranges from 0.03, in cases where no cells are removed from the layer, to 0.06 when cells are removed and/or damaged. Removal or damage was observed for normal loads above around 0.5 mN. Interestingly, the normal loads experienced by a stent *in vivo* are expected to be higher than 0.5 mN, especially in the case of patients with high blood pressure, and therefore cell damage or removal may also play a role in stent displacement. This observation, plus the newly gained information on friction coefficient now permit the modelers to go to the next level of sophistication in their design of effective stents.

Tribology and Lubrication Technology
August 2007, 63(8) p80

Further Reading:

[1] Dunn, A.C., Zaveri, T.D., Keselowsky, B.G. and Sawyer, W.G. (2007). Macroscopic Friction Coefficient Measurements on Living Endothelial Cells, Tribology Letters, 27, pp. 233–238.

Sliding on Cells

Friction measurements on single cells may provide in sights into cell-membrane behavior

There are many medical devices where the sliding of cells against man-made materials is of importance. These include stents, contact lenses and catheters. There is also growing evidence that a number of disease states may be reflected in changes in the measured mechanical properties of cells, including frictional response. This opens the possibility that such measurements could be used diagnostically. Moreover, there are numerous reports that cells react biochemically to mechanical influences — a fact that could be of potential relevance in contact-lens use.

In a previous column [1] we reported on work by the group of Prof. Greg Sawyer from the University of Florida, who had used tribometer measurements to determine the friction coefficients of various materials sliding on aortic endothelial cells [2]. In a more recent study at the same institution, the group of Prof. Scott Perry has measured and analyzed, in collaboration with Sawyer, friction on single corneal epithelial cells in culture medium, by means of an atomic force microscope, incorporating a spherical, colloidal silica tip [3]. The paper is significant for a number of reasons. By using a local-probe method such as AFM, they are able to distinguish the intrinsic behavior of the outer surface of the cell from the collective behavior of a cell layer. This provides mechanical data that could be directly related to the environment of the individual cell, in terms of physiology, disease, or drug stimuli.

175

From a mechanical-engineering standpoint, the paper raises an interesting issue concerning tribological measurements of soft systems such as cells. By employing an elastic foundation ("bed of springs") model, rather than the conventional Hertz-Sneddon approach, the authors were able to extract the load-dependent shear stress at the cell-probe interface. This interface includes the membrane components, the trans-membrane proteins, and the proteins adsorbed from the cell-culture medium onto the probe tip. The shear stress is the consequence of the friction, and the stimulus to which the cell can react biochemically. The ability to measure this for a living cell potentially opens up a new area of tribological-biochemical investigations.

Fig. 1 Scrutiny of the way in which cells react to mechanical influences could prove useful in understanding issues related to contact lens comfort.

The Hertz-Sneddon model makes numerous assumptions, such as a flat, non-frictional contact, infinitesimal deformation, infinite sample thickness and isotropic properties, and is therefore less suitable than a model that takes a more realistic approach to cell-material interactions.

A demonstration that the elastic-foundation approach seems to work for the cells investigated by the Florida groups is that the AFM indentation behavior of the cells can be very well fitted by the model. It is interesting to note that an elastic foundation model predicts a significantly higher (by a factor of five) elastic modulus than that predicted by the Hertz-Sneddon model.

Tribological investigations are sensitive to many mechanical properties and chemical effects. As such they provide a valuable window onto the way in which biological systems react to the stresses of the outside world. The challenges, as these authors have shown, lie both in the measurements and in the data analysis.

Tribology and Lubrication Technology
August 2010, 66(8) p72

Further Reading:
[1] Tysoe, W.T. and Spencer, N.D. (2007). Fenders and Stents, Tribology and Lubrication Technology, 63(8), p80.

[2] Dunn, A.C., Zaveri, T.D., Keselowsky, B.G. and Sawyer, W.G. (2007). Macroscropic Friction Coefficient Measurements on Living Endothelial Cells, Tribology Letters, 27, pp. 233–238.

[3] Straehla, J.P., Limpoco, F.T., Dolgova, N.V., Keselovsky, B.G., Sawyer, W.G. and Perry, S.S. (2010). Nanomechnical Probes of Single Corneal Epithelial Cells: Shear Stress and Elastic Nanomechanical Modulus, Tribology Letters, 38, pp. 107–113.

Keeping up with Contacts

Industrial companies sponsor basic research into contact-lens tribology

In September, STLE-affiliated journal *Tribology Letters'* Editorial Board member, Philippa Cann of Imperial College, London organized the first International Conference on Biotribology (ICoBT), which attracted luminaries from some 38 countries. In addition to the predictably large number of contributions on artificial and natural joints, a significant number of papers concerned the growing areas of skin tribology and contact lenses. Interestingly, many of these studies were industrially sponsored, but nevertheless the results were being communicated to the scientific public.

Several of the contributors to ICoBT have recently published their presented work in *Tribology Letters*. Notable among these is a comprehensive review [1] on skin by Siegfried Derler of the Swiss Federal Laboratories for Materials Science and Technology and Lutz Gerhard of the Technical University of Eindhoven, Netherlands. Skin friction is of great practical importance for applications ranging from the design of sports clothing and swimming-pool walking surfaces, to the avoidance of bed-sores in long-term hospital patients. From the review it is clear that skin friction displays enormous variation, even when taking into account the fact that skin is a highly non-linear viscoelastic material. Much of the variation can be attributed to the degree of skin hydration, but overall it can be stated that the adhesion contribution to skin friction dominates over that from deformation. The review contains a valuable compilation of skin-friction data from many researchers, measured on several different areas of the body and against a wide variety of countersurfaces.

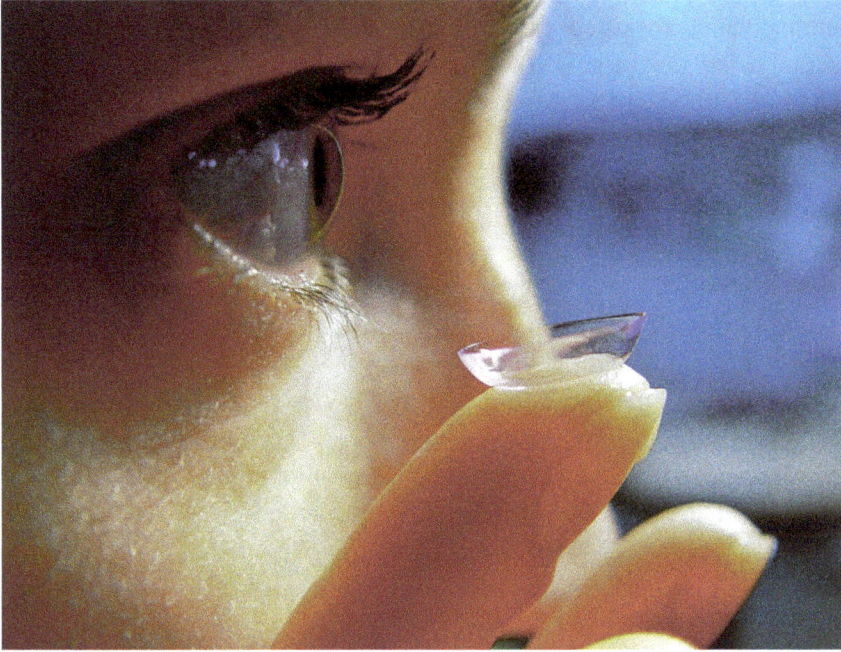

Fig. 1 "Contact Lens Ayala" by לטומתיא - Own work. Licensed under Creative Commons Attribution 3.0 via Wikimedia Commons

We reported in August 2010 about efforts at the University of Florida to measure friction on corneal epithelial cells [2]. Two recent papers in *Tribology Letters*, again by the Florida group [3] but also from the group of Samuele Tosatti at the Swiss company SuSoS, together with their collaborators at the ETH Zurich [4], have focused on the contact-lens countersurface. The friction of a contact lens against the eyelid is one of the central factors influencing comfort for the wearer, but contact lenses have proven a challenging area in which to conduct tribological research, partially because the contact pressures involved in the application are so low (3-5 kPa) and thus the availability of test equipment in this range has been limited until recently. Test methods were central to the paper of the Tosatti group, who made efforts to ensure that both the countersurface and the test fluid were physiologically relevant. To this end, they employed a glass countersurface that had been modified first by hydrophobization, and then by adsorption of mucin, in order to mimic the inner surface of an eyelid. The lubricant solution, which was to mimic tears, consisted of a buffered salt solution containing human serum plus the protein lysozyme. Friction results on a number of commercially available contact lenses revealed that those containing the hydrophilic polymer poly(vinyl pyrrolidone) (PVP) showed the lowest friction coefficients (generally less than 0.05), while those without PVP showed μ values in the range 0.1 to 0.6.

The group of Scott Perry at the University of Florida have focused on surface composition (measured by x-ray photoelectron spectroscopy) and friction (measured by atomic force microscopy) of several contact lenses based on silicone hydrogels. Additionally, they investigated the effect of contact lens pretreatment in a solution of a poly (ethylene oxide)-poly (butylene oxide) (EO-BO) copolymer, which served as a model additive in a contact-lens storage solution. Surface hydrophilicity is clearly of importance in imparting low friction to contact lenses. Of the three types of lenses examined, two were hydrophilic by virtue of a plasma-treatment step during their manufacturing process. The other lens was not plasma-treated, but contained PVP as an internal wetting agent. As in the Tosatti study, the PVP-containing lens showed the lowest friction coefficient. Interestingly, only the plasma-treated lenses showed significant adsorption of the EO-BO copolymer at the lens surface, and thus only these lenses showed any reduction in friction upon treatment with polymer. The results show not only that lens solutions can impact lubricity of lenses, but also that the extent of this effect can depend on the type of lens employed.

Clearly contact lenses bring enormous benefits to millions of wearers, and any improvement in their effectiveness or comfort can be extremely valuable. The increasing desire of the lens manufacturers to sponsor basic research in understanding lens materials chemistry and tribology is a significant development, and their willingness to allow its publication in the public domain is to be applauded.

Tribology and Lubrication Technology
April 2012, 68(4) p80

Further Reading:

[1] Derler, S. and Gerhardt, L.-C. (2012). Tribology of Skin: Review and Analysis of Experimental Results for the Friction Coefficient of Human Skin, Tribology Letters, 45, pp. 1–27.

[2]Tysoe, W.T. and Spencer, N.D. (2010). Sliding on Cells, Tribology and Lubrication Technology, 66, pp. 72.

[3] Huo, Y., Rudy, A., Wang, A.,Ketelson, H. and Perry, S.S. (2012). Impact of Ethylene Oxide Butylene Oxide Copolymers on the Composition and Friction of Silicone Hydrogel Surfaces, Tribology Letters, 45, pp. 505–514.

[4] Roba, M., Duncan, E.G., Hill, G.A., Spencer, N.D. and Tosatti, S.G.P. (2011). Friction Measurements on Contact Lenses in Their Operating Environment, Tribology Letters, 44, pp. 387–397.

Hip Surface Science

A new paper gives us a better understanding of how articular joints are lubricated.

One of the most astonishing tribological systems in nature is that found in our hips and other articular joints. The joints show exceedingly low sliding friction, contain their own lubricant-producing machinery, and in most cases last for many decades without significant wear. Obviously there are good reasons for trying to understand how these remarkable tribosystems work, not least because our aging population is reaching the end of the useful life of their hip or knee joints in ever-increasing numbers.

Over the years there has been a multitude of different mechanisms invoked to describe the lubrication of articular joints. These have included boundary lubrication, elastohydrodynamic lubrication, microelastohydrodynamic lubrication, squeeze-film lubrication, "weeping" lubrication, "boosted" lubrication, electrostatic lubrication, biphasic lubrication, brush lubrication, and gel lubrication to name but a few! In fact, many of these mechanisms may be involved to a greater or lesser extent, or in combination, under different conditions of joint use. Nevertheless, it is clear that in many situations, joints are operating from a standstill, and therefore some form of boundary lubrication will certainly be playing a role, and the cartilage surface properties will be highly relevant.

In her recent review article published in *Tribology Letters*, Rowena Crockett, of the Swiss Federal Laboratories for Materials Testing and

Research in Dübendorf, Switzerland, has focused specifically on boundary lubrication of articular joints [1]. In particular, she handles the delicate issue of the cartilage surface. This is a trickier topic to investigate than might be expected since studies of explanted cartilage have been prone to artifacts and the near-surface region is a highly complex, layered structure. Almost all investigators agree on the components present, however. These include *phospholipids* (the building blocks of membranes), *hyaluronan* (a charged, high-molecular-weight sugar chain), *lubricin* (a protein with sugar side-chains, or *glycoprotein*) and *collagen*, the main protein of connective tissue.

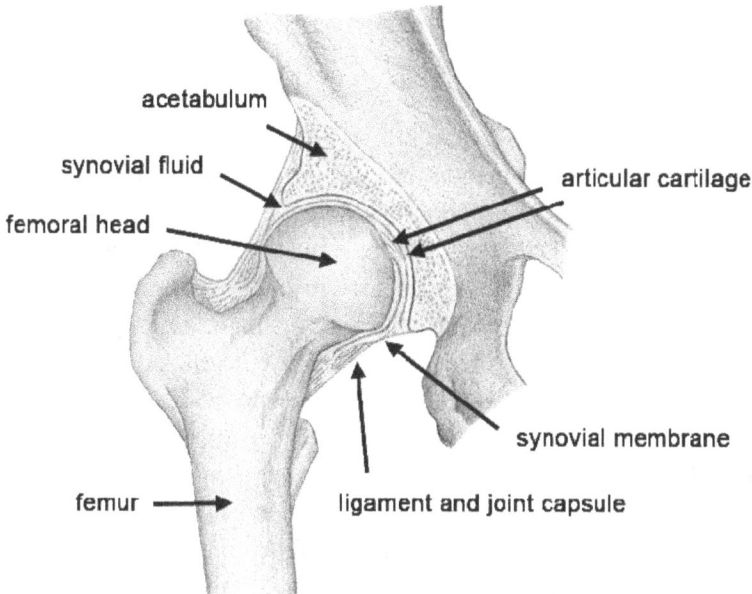

Fig. 1 Schematic diagram of natural hip joint (kindly provided by Sasa Vranjkovic, Empa, Switzerland).

Working through the many studies that have been published over the last few decades, Crockett was able to put together a view of the cartilage surface that seems consistent with much of the evidence, even though this has previously been interpreted in many different and conflicting ways. Starting inside the bulk cartilage, the cartilage cells, or *chondrocytes*, are embedded in a sea of collagen and supramolecular structures containing proteins, hyaluronan, and glycoproteins, known as *aggregate*. This bulk structure starts near the bone and goes out to the near-surface region, but we are not at the sliding or *articular* surface yet. On top of this collagen-rich phase is a hydrophilic layer, or possibly layers, probably consisting of lubricin, and very closely related molecules, complexed with hyaluronan. This sugary structure forms the articular surface. Finally, on top of the articular surface is a dilute, gel-like network of hyaluronan and lipids, which allows facile diffusion of molecules in and out of the cartilage below. Lipids have been shown to dramatically reduce the viscosity of hyaluronan, so this may be the origin of the lubricious, low-shear-strength layer present on top of the cartilage surface.

This outer, gel-like surface (or "superficial layer") of the cartilage is delicate and highly susceptible to drying out and becoming hydrophobic — a phenomenon that has previously led to the hypothesis that cartilage itself is hydrophobic, but as we have seen, the situation is far more complex than this. Methods that have been used to image cartilage surfaces using electron microscopy generally remove the superficial layer, the articular surface, or both, to expose the bulk cartilage. This has also led to inferences based on bulk cartilage morphology, which probably have little to do with the actual sliding surface properties.

Cartilage clearly has a far more complex structure than we are used to in engineering systems, and it works extremely well. Let us hope that a greater understanding of its structure and function will lead to new ways to deal with diseases of the joints, and possibly even new directions for the design of materials for low-friction mechanical systems.

Tribology and Lubrication Technology
August 2009, 65(8) p56

Further Reading:

[1] Crockett, R. (2009). Boundary Lubrication in Natural Articular Joints, Tribology Letters, 35, pp. 77–84.

How Many Licks? The Lollipop and Biotribocorrosion

How the mind-bendingly complex world of biotribocorrosion can be accessed by basic tribological principles, careful metrology, statistical modeling, and an army of lollipop-lickers.

Readers of this column do not need to be told that tribology is a complex, multidisciplinary topic. Biotribology and tribocorrosion bring with them even greater levels of complexity, while biotribocorrosion may seem like a lost cause when it comes to quantitative understanding and prediction. Nevertheless, the topic of biotribocorrosion is of great importance in applications ranging from implants to marine engineering.

A refreshingly original, potentially important, and not to say thoroughly entertaining article on a well-defined biotribocorrosion problem was recently published in *Tribology Letters* by a group from the University of Florida led by Professor Greg Sawyer [1]. The paper posed the question "How many licks does it take to reach the center of a lollipop" — specifically the soft core of a Tootsie-Pop®. The researchers approached the problem armed with classical tribological methods, state-of-the-art optical metrology equipment, Monte Carlo statistical modeling techniques, and dozens of volunteer lickers from a local high school and the STLE annual meeting.

One of the issues that has made quantitative studies of (bio)tribocorrosion challenging has been the dual processes of corrosion and wear, which are on the one hand chemical and time-dependent and on the other hand mechanically driven and cycle dependent. To make things more challenging, these processes are coupled to each other in a complex way. In this study, the situation was simplified by lumping the corrosion and wear aspects together in a single, cyclic volumetric material removal parameter, supported by the statistical data from the 58 lickers. The experimental data were then fed into Monte Carlo simulations, varying the input parameters to a simple wear equation, based on the variability observed in the measured wear rates.

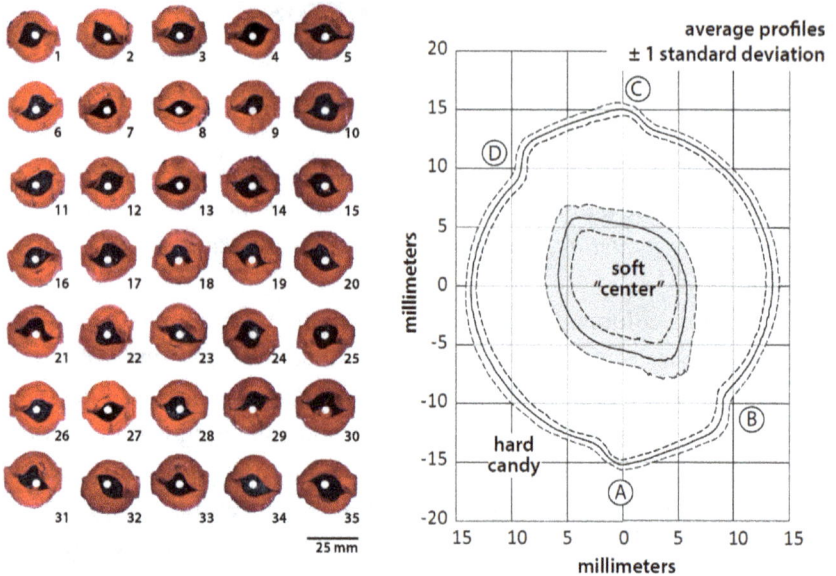

Fig. 1 (Left) 35 equatorial cross-sections of randomly selected Tootsie Pops® illustrating the variation in shape and size, especially of the soft Tootsie Roll® centers. (Right) The composite average Tootsie Pop® outline. By kind permission of Springer Science+ Business Media from Reference [1].

First, the lollipops had to be characterized, by weighing and shape analysis. The latter involved cutting them equatorially in a band saw, polishing them, and subjecting them to optical metrology (Figure 1a). In this way an "average" Tootsie Pop® could be defined (Figure 1b). Also, a number of control experiments had to be performed, such as determining the mass loss when licking with a dry tongue (negligible removal rate), or dissolution in a water bath or in a mouth with no movement (around 7-9 mg/s, depending on degree of agitation). In contrast, typical mass-loss rates during licking were in excess of 17 mg/s, clearly indicating the synergy between wear and dissolution during the licking process.

The sliding partner during licking, of course, was the tongue or palate, and typically has an R_a of around 33 μm, a modulus of about 15 kPa, and exerts a normal force of around 3 N (corresponding to a few kPa) during licking. Two clear modes of lollipop licking were observed among the volunteers: single-sided (tongue only, 17 mg per lick) and full-surface (tongue and palate, 62 mg per lick). The licking experiments themselves involved weighing after every 10 licks, which had to be performed in a consistent manner. From these data, the volumetric removal per lick could be determined and hence a lumped-parameter wear rate.

5,000 Monte Carlo simulations of the evolution of surface geometry were carried out, based on the range of input parameters determined from the experiments. Interestingly for the first 150 cycles, the average wear depth was found to vary little between the two licking styles, despite the large differences in volumetric wear rates. This is presumably due to the large difference in contact area between the two styles.

The pressure regime and chemical environment in this study is not so dissimilar from that of many other biological environments where biotribo corrosion is a concern. The statistical approach, coupled with the lumping of processes with intrinsically different evolutions over time is highly relevant to the treatment of problems with greater biomedical significance than lollipops. Amusing though this study may be, it could be serve as a valuable stimulus to further advances in biotribocorrosion.

By the way, the answer is 130 licks to the center.

Tribology and Lubrication Technology
December 2014, 70(12) p104

Further Reading:

[1] Rowe, K.G., Harris, K.L., Schulze, K.D., Marshall, S.L., Pitenis, A.A., Urueña, J.M., Niemi, S.R., Bennett, A.I., Dunn, A.C., Angelini, T.E. and Sawyer, W.G. (2014). Lessons from the Lollipop: Biotribology, Tribocorrosion, and Irregular Surfaces, Tribology Letters, 56, pp. 273–80.

Sliding Soft Surfaces

Measuring the speed dependence of hydrogel friction helps us understand natural lubricated systems

Sliding interfaces in humans and animals, either in articulating joints or between the eyelid and the eye, are required to undergo thousands of cycles per day for many years. The contacting materials invariably consist of nanoporous, hydrophilic polymer networks that are permeable to water. Synthetic networks of hydrophilic polymer chains, such as those made from polyacrylamide, have been used as models to study the tribological interfaces encountered in biological systems. Most of these experiments measured the friction between the hydrogel and hard, impenetrable counterfaces, while biological interfaces are invariably made from identical "Gemini" contacts (Fig. 1). The group of Professor Greg Sawyer along with colleague Thomas Angelini from the University of Florida remedied this deficiency by comparing the friction of a Gemini interface consisting of a polyacrylamide, hydrogel ball sliding against a hydrogel substrate [1]. To gain a deeper understanding of the friction behavior, they compared the results with those for a glass ball sliding against a hydrogel substrate and a hydrogel ball sliding against a glass substrate. In all cases, the hydrogels were carefully hydrated with ultrapure water.

The results showed some remarkable differences in behavior for the three contact conditions. For a migrating contact of a glass ball sliding against a hydrogel, where the position of the hydrogel contact changes as the ball slides, the friction coefficient was found to decrease rapidly with sliding speed. For a stationary contact of a hydrogel ball sliding against glass, a much weaker sliding-speed dependence was found, while the Gemini contact had very low friction (μ< about 0.06), with no detectable velocity dependence.

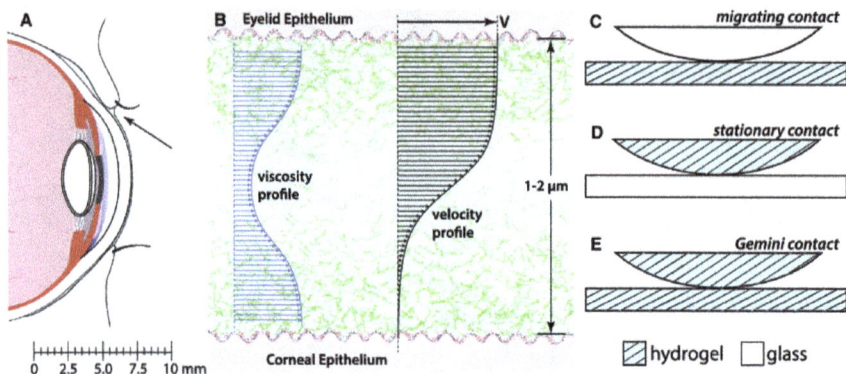

Fig. 1 A The eye is an exquisite system of lubrication comprising the epithelia and tear film. B The tear film composition is graded, with a higher concentration of mucins and glycoproteins near the epithelia. This schematic illustrates how the local sliding velocity between the eyelid and corneal epithelia drops off severely closer to the surfaces due to the high-viscosity mucinous gels. It is this system of self-mated hydrated gel lubrication that motivates this work using synthetic gels in aqueous lubrication. C–E Each contacting condition explored in the work is represented here as a schematic showing the hydrogel and glass components in the following order: migrating contact with glass probe and hydrogel flat; stationary contact with a hydrogel probe and glass flat; and self-mated, or Gemini contact, where both surfaces are hydrogel. By kind permission of Springer Science+Business Media from Reference [1].

Insights into this behavior came from measuring the friction as a function of the time of the experiment for the migrating and stationary contacts, where the data appeared to scale as time to the power of ~1/3. To understand this behavior, Sawyer used a law proposed by a French hydrologist, Henry D'Arcy in the middle of the 19th century to study the permeation of water through sand, which successfully describes fluid flow through porous media, such as hydrogels or sponges, as well as sand or clay. The law states that the fluid flow rate is proportional to the pressure drop across the material — in the case of the friction experiments, the contact pressure. Making some simplifying assumptions, Sawyer found that the contact area varied as $t^{1/3}$, rationalizing the observed friction behavior if the friction force scales with contact area.

This now sets the stage for understanding the velocity dependence of the friction for the various contacts. In the migrating, glass-ball-on-hydrogel contact, where the friction coefficient depends strongly on speed, larger velocities leave less time for the contact to grow, producing a lower contact area, and hence lower friction. In contrast, the contact in the static, hydrogel-on-glass interface is at one spot on the gel so it is under persistent load. Thus the contact area does not vary strongly with speed, producing a weak velocity dependence.

The Gemini, gel-on-gel situation is a combination of both conditions, in which the probe contact does not change, while that of the substrate does. However, since the friction remains constant and low for all sliding velocities, its behavior is not a simple superposition of the two asymmetric contacts. It is suggested that such low friction arises because both of the surfaces at the Gemini interface are persistently coated with water, minimizing the possibility of polymer-polymer interactions. The polymers are also themselves hydrated, further reducing interactions.

It appears that such hydrogel interfaces can help us gain insight into the fundamental processes occurring in biological lubricated systems. Indeed, extending these ideas to cartilage, which is about two orders of magnitude less permeable than hydrogels, suggests that the fluid permeation timescale for hip or knee joints is several hours. In other words, standing for long periods of time does not necessarily cause high startup friction.

Tribology and Lubrication Technology
June 2014, 70(6) p96

Further Reading:

[1] Dunn, A.C., Sawyer, W.G. and Angelini, T.E. (2014). Gemini Interfaces in Aqueous Lubrication with Hydrogels, Tribology Letters, 54, pp. 59–66.

The Contact Conundrum

Understanding the area of contact between two solids is fundamental to tribology and many books have been written on the subject. At least for smooth elastic surfaces, the ideas of Hertz, that he formulated over a single Christmas holiday in the 19th Century, have been used ever since as the basis for calculation contact areas. However, the real engineering situation is much more complex, described by Archard as being due to the contact between rough surfaces having "protuberances on protuberances on protuberances".

Mathematical theories have been developed to describe such shapes, which are knows as fractals. The use of these theories to study the contact between two rough surfaces was discussed in our first article in this area, *Searching for the fractal truth* in February 2005, and appeared to provide a definitive answer to this problem.

However, this appeared not to be the case, even for simple elastic contacts, when the detailed atomic structures of the contacting surfaces were taken into account (*The continuing contact conundrum*, October 2005).

The nature of the contact has also been found to influence the commonly observed stick-slip behavior that produces squeaking doors, the sound of a violin, and even the shaking during an earthquake. It has been observed that the extent of slick slip depends on the contact; may small contacting protuberances tend to give less stick slip, while a few large ones give more. A simple model for this was discussed in *Split or squeak?* in December 2014 that showed that a single large contact did indeed lead to large stick-slip behavior, which was reduced by creating two equivalent contacts, and disappeared almost completely for a contacting interface in which there tens of contacts.

Much of the uncertainty in this field arose from the experimental difficulties of actually seeing what was happening at the contact between two rough surfaces. This problem was solved in *The contact conundrum cracked?*, February 2007, by the simple approach of using rough, transparent, Plexiglas blocks. The trick was to use the fact that the refractive indices of the blocks matched when they were in contact so that light passed through them, while light was reflected at the Plexiglas air interface in the non-contacting area. This allowed the real contact area to be measured, at least for Plexiglas. This trick has also been used to study what happened when the two surfaces were sheared, as we discussed in the chapter on *Fundamentals of Friction and Damage*.

The waters were, however, muddied, as discussed in *Contact conundrum conquered?*, June 2009, which reported on a study of the effects on contact area of short- and long-range interactions. Here it was found that, when only sort-range interactions were present, not all of the atoms in the asperities are really in contact.

Apparently the contact conundrum continues!

Searching for the Fractal Truth

The fractal nature of real, engineered surfaces is used to calculate the real contact area between bodies

Early in the last century, Lewis F. Richardson (Fig. 1), who was an ardent pacifist, was trying to discover why countries went to war. One of his ideas was that the tendency of two countries to go to war was proportional to the length of their common border.

To try to test this idea, he measured the length of the coastline of the British Isles using various maps, but obtained completely different results for different scale maps. He discovered that the lengths that he measured were not in error, but depended on the length of the "ruler" that was used to make the measurement; the shorter the length of the ruler, the longer the measured length of the coastline.

He also found that there was a simple power-law relationship between the measured length of the coastline and the length of the ruler used to measure it. This behavior has to do with the inherent "roughness" of the coastline so that as the ruler is shortened, more and more detail becomes visible that was not seen previously. Such shapes are known as "fractals" and have the property that they look similar over a wide range of length scales. These are very common in nature and examples include shapes of leaves, snowflakes, mountains, and coastlines.

The tribological relevance is, of course, that rough, engineered surfaces are also fractal. This was pointed out many years ago by Archard, who emphasized that rough surfaces did not have asperities of a single size, but consisted of "protuberances upon protuberances upon protuberances" (Fig. 2).

Fig. 1. Lewis F. Richardson, 1881–1952.

A central problem in tribology has been to relate the real contact area between surfaces to the force acting on them. Amonton's law implies that the real contact area is proportional to the load, while, for an elastic Hertzian contact, it varies as the load to the 2/3 power [1].

In the '70s, Bush and co-workers addressed this question by examining the contact of a surface with a distribution of asperity sizes. Gratifyingly, it was found that the contact area depended linearly on the applied load. More recently, the properties of truly fractal surfaces were investigated by Bo Persson, in Forschungszentrum Jülich, Germany who also found that the real area of contact of a fractal surface was proportional to the applied load [2]. Interestingly, both the Bush and Persson theories arrived at the same rather elegant conclusion that the real contact area per unit applied load was inversely proportional to both the reduced Young's modulus and the root-mean square slope of the surface.

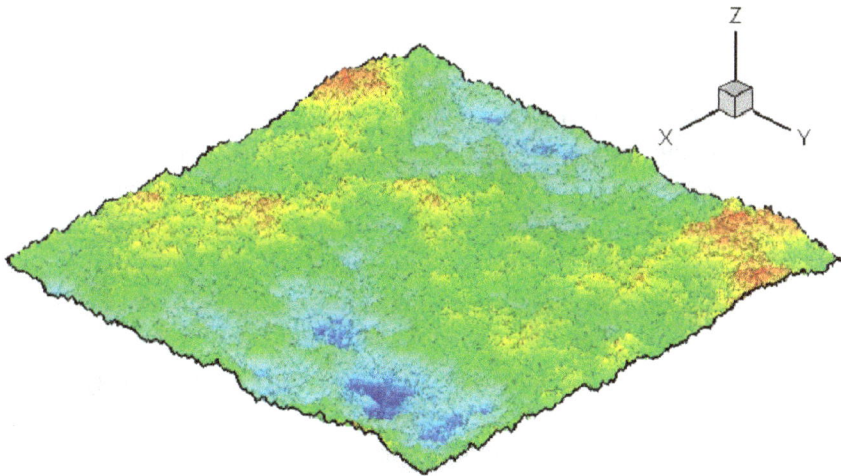

Fig. 2. Self-affine fractal surface image. Heights are magnified by a factor of 10 to make the roughness visible, and the color varies from dark (blue) to light (red) with increasing height. Reprinted with kind permission from Reference [2]. Copyright (2002) by the American Physical Society.

The difference between the two theories is the value of the proportionality constant; the Bush theory yielding ~2.5, while Persson's fractal theory gave ~1.6. More recently, Mark Robbins and co-workers, at Johns Hopkins University have carried out finite-element analyses of fractal surfaces and confirmed the general simple conclusions of the analytical models [3]. They also found that the proportionality constant depended rather weakly on parameters such as surface roughness and Poisson's ratio, but always lay between the limits of the two models.

They suggested that a value of ~2.2 should correctly predict the real contact area to within about 10%. These exciting results for realistic fractal surfaces promise to provide answers to one of the central questions in tribology — *what is the true contact area between engineered surfaces*?

Tribology and Lubrication Technology
February 2005, 61(2) p56

Further Reading:

[1] Tysoe, W.T. and Spencer, N.D. (2004). Why Does Amontons' Law Work so Well?, Tribology and Lubrication Technology, 60(8), pp. 56.

[2] Persson, B.N.J., Bucher, F. and Chiaia, B. (2002). Elastic Contact Between Randomly Rough Surfaces: Comparison of Theory with Numerical Results, Physical Review B, 65, pp. 184106.

[3] Hyun, S., Pei, L., Molinari, J.-F. and Robbins, M.O. (2004). Finite-element Analysis of Contact Between Elastic Self-Affine Surfaces, Physical Review B, 70, pp. 026117.

The Continuing Contact Conundrum

Molecular dynamics simulations of various tip structures are used to test contact mechanics theories at the nano scale

The nature of the contact between a cylinder or sphere and a flat surface is one of the central issues in tribology. Heinrich Hertz first solved this problem over 120 years ago for macroscopic elastic contacts, by analogy with electrostatic theory [1]. As we have discussed in a previous column, real engineered surfaces are rough at the microscopic scale and can be modeled as fractals consisting of, in the words of Archard, "protuberances upon protuberances upon protuberances" [2].

In this case, some of the contacting asperities are of atomic dimensions and with the advent of atomic force microscopy (AFM), it has become possible to measure the contact and frictional behavior of such nanoscale asperities. The question then naturally arises "How well does continuum mechanics — the Hertzian theory — describe such nanoscale contacts?" Mark Robbins and graduate student Binquan Luan of Johns Hopkins University have recently addressed this question [3].

They used molecular-dynamics simulations of nanometer-scale cylinders of the same diameter, but constructed in different ways, placed in contact with a flat, crystalline surface (Fig. 1). First, a cylinder-flat contact was made from a slab of a face-centered cubic material that was either bent so that the resulting lattice spacing was the same as or an irrational multiple of the substrate lattice spacing (so-called commensurate or incommensurate, respectively).

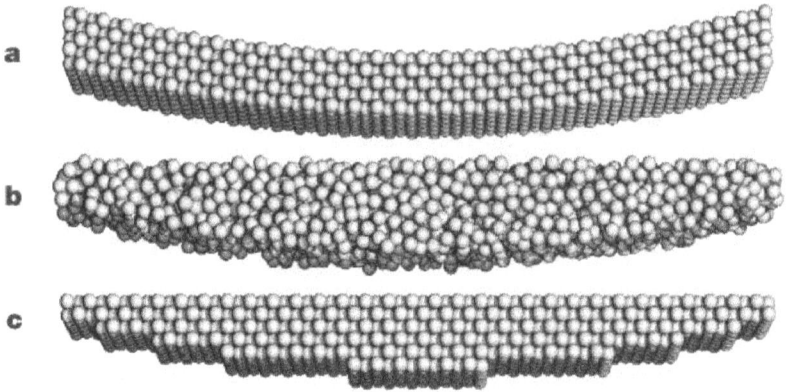

Fig. 1 Snapshots of atoms near the bottoms of cylindrical surfaces of average radius R = 100 sigma formed by bending a crystalline slab (a), or cutting an amorphous (b) or crystalline solid (c). The steps in c are not a unique function of R, which leads to further variations in the behavior of such tips. The tips are pressed down onto a horizontal elastic substrate. Periodic boundary conditions are applied along the axis of the cylinder, which runs into the page. Reprinted by kind permission from Macmillan Publishers Ltd: Nature (Reference [3]), copyright (2005).

Second, the contact was constructed by cutting an amorphous material to make a randomly rough cylinder, and finally a single crystal of the material was cut to form a cylinder, thereby introducing steps. In all cases, the roughness of the tip was about the size of the atoms but the different ways of forming the cylinders resulted in different surface topographies and, in all cases, the average radius was about 30 nanometers — roughly the size of an AFM tip.

Robbins and Luan found that the continuum model worked surprisingly well for some quantities, even for these nano-sized asperities where, for example, the contact penetration into the substrate was very close to the predictions of Hertz theory (Fig. 2).

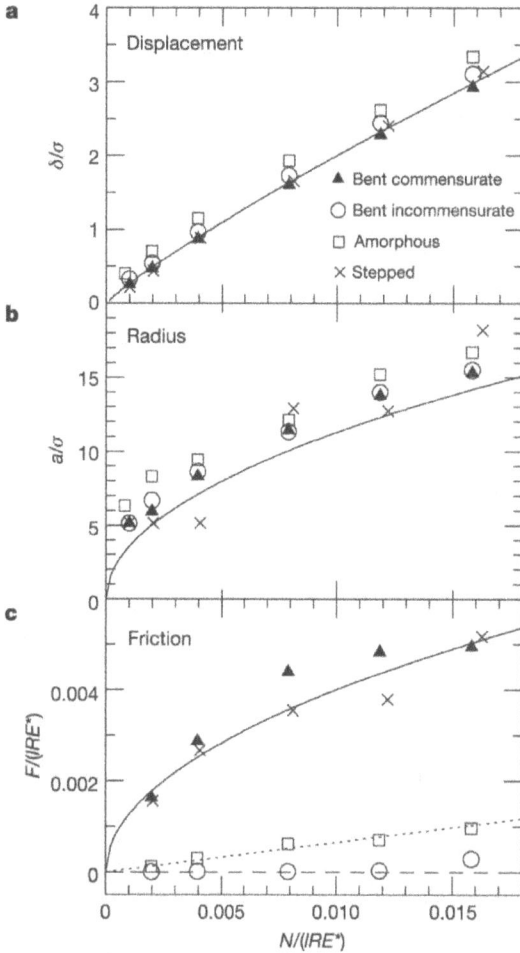

Fig. 2 Dimensionless plots of normal displacement δ (a), contact radius a (b) and static friction F (c) *versus* normal force N for the given cylinder surfaces (symbols). For stepped surfaces, a increases in discrete jumps. Here $R = 100\ \sigma$, the zero of δ corresponds to the first non-zero force, and a is half the range over which the force on substrate atoms is non-zero. Solid lines show continuum predictions, with the assumption that F is proportional to area in c. The broken lines in c are linear fits. Reprinted by kind permission from Macmillan Publishers Ltd: Nature (Reference [3]), copyright (2005).

The contact area was reproduced rather less well by the theory, since the atomic roughness spreads the pressure over a larger area than would be expected for a surface that is completely smooth and may lead to errors up to 100% for small loads. As one would expect, this error decreases as the tip or the applied load become larger. The structure of the asperity had a drastic effect on the contact-pressure distribution, however, with stepped and amorphous tips showing large fluctuations in the local pressure.

The authors also calculated the lateral force required to initiate sliding and here they found the largest differences. Tips whose lattice spacing was commensurate with the substrate show the largest friction since atoms in the tip can lock into local registry with the substrate. Correspondingly, they found that both amorphous tips and those formed by bending the material so that it was incommensurate with the substrate showed the lowest friction. Indeed, the incommensurate surface showed almost no friction. Interestingly, the relatively low-friction contacts showed lateral forces that varied linearly with normal load, but not with contact area.

While these intriguing results emphasize the difficulties of precisely determining contact areas for even apparently well-characterized and simple systems, and emphasize the importance not only of accurately knowing the geometry but also the surface structure of the contact, they do suggest that it might be possible to tailor interfaces with unique tribological properties and perhaps very low friction by carefully controlling the atomic structure of their surfaces.

Tribology and Lubrication Technology
October 2005, 61(10) p64

Further Reading:

[1] Tysoe, W.T. and Spencer, N.D. (2004). Why does Amontons' Law Work so Well?, Tribology and Lubrication Technology, 60(8), p64.

[2] Tysoe, W.T. and Spencer, N.D. (2004). Searching for the Fractal Truth, Tribology and Lubrication Technology, 61(2), p56.

[3] Luan, B. and Robbins, M.O. (2005). The Breakdown of Continuum Models for Mechanical Contacts, Nature, 435, pp. 929–932.

The Contact Conundrum
Cracked?

The real area of contact between Plexiglas slabs was measured using a laser showing that the real area of contact was only a few percent of the apparent contact area

We have previously discussed the difficulties involved with the precise calculation of contact areas at a realistic tribological interface [1,2]. Most of the problems with validating any theory arise from the difficulty in precisely measuring real contact areas at the solid-solid interface between rough surfaces.

This issue has recently been elegantly addressed by the group of Professor Jay Fineberg at the Hebrew University of Jerusalem by using an interface constructed from a block of Plexiglas with a root-mean-square (RMS) roughness of 0.1 to 15 μm that is placed in contact with a much smoother glass or Plexiglas surface (with an RMS roughness of ~100 nm) [3].

The interface is then illuminated by a laser beam where the incidence angle is selected to be larger than the critical angle for the Plexiglas-air interface, so that all of the light is reflected. However, when the two surfaces are brought into contact, light is transmitted through the asperity-asperity contacts at the interface so that the transmitted light intensity is, to a very good approximation, directly proportional to the real area of contact, thus providing a direct measure of the real contact area (Fig. 1).

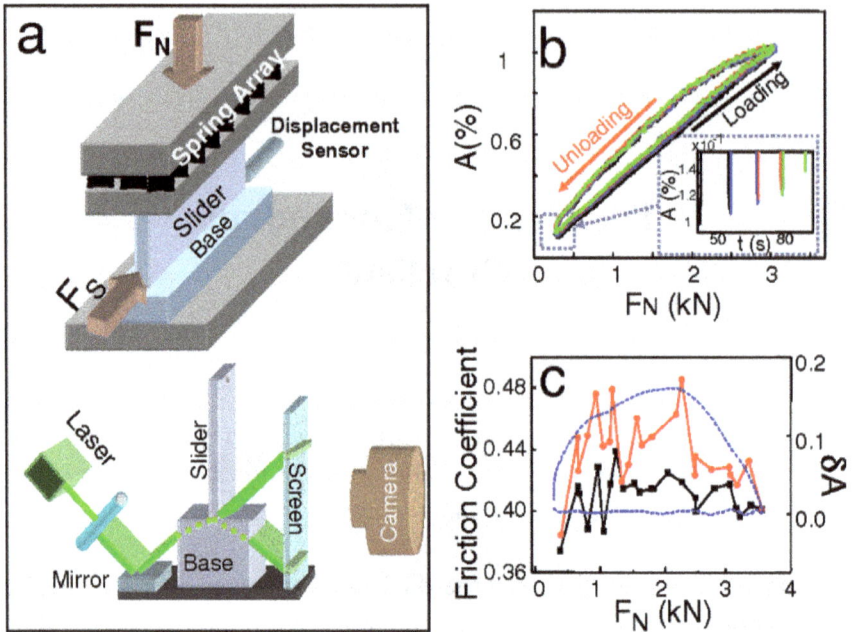

Fig. 1: (a) A schematic view of the experimental system. (top) Transparent slider and base blocks are loaded through a soft 40-spring array. Displacement is measured at both the leading and trailing edges (where F_S is applied) of the slider. (bottom) A laser sheet illuminates the interface though the base. Both the transmitted and reflected beams are imaged onto a fast camera. (b) The actual area of contact, A (in % of nominal contact area), vs F_N for 4 successive loading or unloading cycles. The samples were not separated between cycles and a minimum load of 270 N was retained. (inset) A, as a function of time at the lowest values of F_N, shows negligible contact area accumulation between consecutive cycles. (c) Static friction coefficient, μ_s directly measured upon loading (▪) and unloading (●) over an F_N cycle, as in (b). The dashed line indicates the difference in contact area between loading and unloading, δA, as derived from (b). In (b) and (c) the slider and base were of PMMA, with a nominal contact area of 150×6 mm. Reprinted with kind permission from Reference [3]. Copyright (2006) by the American Physical Society.

The authors found that when they loaded the surface, the real contact area between contacting asperities was only a few percent of the nominal contact area, and varied almost exactly linearly with applied load, as is generally expected. Moreover, the measured contact areas agreed well with the values calculated using a Gaussian distribution of contact points with a density and radius of curvature that were consistent with the specimens used for the experiments. Furthermore, the friction coefficient (of ~0.4) was independent of applied load.

Surprising results were found, however, during unloading. In this case, the contact area, when plotted as a function of applied load, did not follow the same path as during loading, but was consistently higher — although it reduced to zero as the load disappeared. That is, the loading-unloading curve showed hysteretic behavior.

Moreover, this hysteretic behavior was not only found in the first cycle but occurred repeatedly and reproducibly as the system was loaded and unloaded. Measurements of the friction coefficient during unloading showed consistently higher values, scaling rather well with the contact area.

Being able to measure real contact areas allowed the average contact pressure to be measured and revealed that the contacts predominantly deformed plastically. This would be expected to produce hysteresis, but only in the first cycle, as the surface deformed, and not in subsequent cycles, when the deformation should be complete as long as the maximum normal force is not increased.

In fact, when unloading and loading cycles were applied to different maximum values, when the area and load data were normalized to the maximum contact areas and applied loads, all the data collapsed onto a single curve.

Fineberg and co-workers suggested that such repetitive hysteresis cycles would be found if renewal of microcontacts were to occur between cycles. It was proposed that this could happen by Poisson expansion (the squeezing out of the block) as the sample was loaded.

This would cause the asperities to move laterally, as the renewal of asperity contacts would only occur if the lateral displacements were larger than the size of a single contacting asperity. To test this conjecture, the Poisson expansion was measured directly using displacement sensors placed on opposite sides of the sample, where expansions of tens of micrometers were measured. This theory further predicts that if one of the surfaces is made extremely smooth so that when the asperity contacts break and reform, they will no longer result in any change in contact area, so that the hysteresis should disappear. Satisfyingly, this is exactly what was found.

Tribology and Lubrication Technology
February 2007, 63(2) p64

Further Reading:

[1] Tysoe, W.T. and Spencer, N.D. (2005). Searching for the Fractal Truth, Tribology and Lubrication Technology, 61(2), p56.

[2] Tysoe, W.T. and Spencer, N.D. (2005). The Continuing Contact Conundrum, Tribology and Lubrication Technology, 61(10), p64.

[3] Rubenstein, S.M., Cohen, G. and Fineberg, J. (2006). Contact Area Measurements Reveal Loading-History Dependence of Static Friction, Physical Review Letters, 96, pp. 256103.

The Contact Conundrum
Conquered?

A new set of simulations of the contact in an atomic force microscope are challenging our thinking

One of the central issues in tribology is to correctly determine the true area of contact between two surfaces. In large-scale contacts, the real contact area is made up of a number of contacting asperities [1]. One of the central ideas behind using atomic force microscopy (AFM) to study friction is that the sharp tips used in this technique model a single-asperity contact. Moreover, it was thought that the contact mechanics could be described using analytical theories developed by Hertz [2], which showed that the area of contact was proportional to $(\text{Load})^{2/3}$. This theory has been modified to include the effects of adhesion between the contacting surfaces, but still yields analytical formulae with sublinear dependences of contact area on load.

In this case, if the lateral force is proportional to the real contact area, where the proportionality constant is defined as the interfacial shear strength, this will result in the lateral (frictional) force being a sublinear function of load, and this has indeed been observed in a number of AFM experiments. However, in some cases, deviations have been observed from such simple elastic theories, where the frictional force is found to be proportional to the load or even where the two quantities were almost independent of each other. This raises the question whether such analytical continuum contact models can really describe behavior at the nanoscale.

207

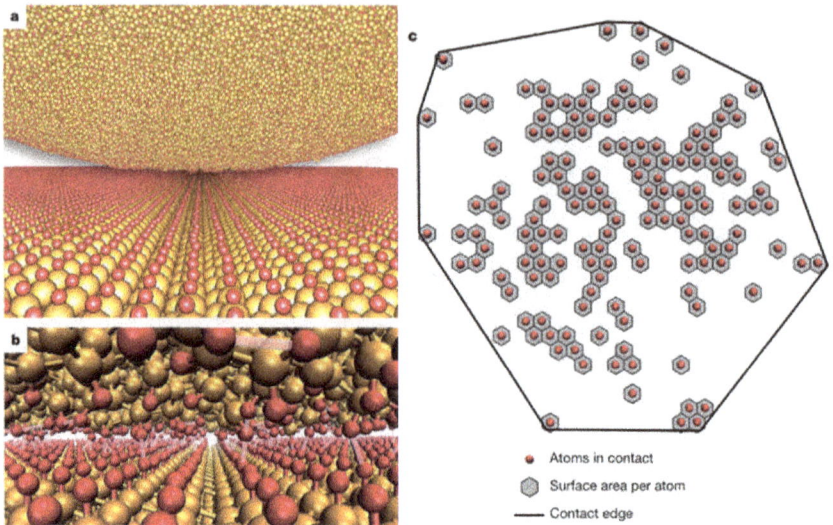

Fig. 1 **a,** Far view, showing contact geometry. Golden and red atoms correspond to C and H, respectively. **b,** Close view. Solid red and golden sticks represent covalent bonds. Translucent pink sticks represent repulsive interactions. **c,** Contact area definitions. Red circles represent sample atoms within the range of chemical interactions from tip atoms. Contact area per atom A_{at} is represented by grey hexagons. Real contact area A_{real} is the sum of the areas of hexagons. The contact area A_{asp} of an asperity is enclosed by the edge (solid line) of the contact zone. Reprinted by kind permission from Macmillan Publishers Ltd: Nature (Reference [3]), copyright (2009).

This issue was addressed by the group of Professor Izabela Szlufarska at the University of Wisconsin-Madison by using large-scale molecular-dynamics simulations to calculate the frictional force at the nanoscale [3]. These simulations were performed for hydrogen-terminated, amorphous-carbon tips and a diamond substrate that allowed both the tip and the substrate to deform (Fig. 1). The interaction potential included both short-range, chemical interactions and longer-range Van der Waals' forces. A key issue was to determine the real contact area correctly, and this was done by calculating the number of chemically interacting atoms in the contact, multiplied by the average surface area occupied per atom.

They first performed simulations in which they turned off the long-range interactions. In this case, they found a sublinear dependence of asperity contact area on applied load, in accord with the predictions from contact mechanics, but a linear dependence of lateral force on load. Furthermore, the composite Young's modulus calculated using the simulations was substantially less than that predicted by continuum mechanics.

The authors argue that this apparent breakdown of single-asperity theories arises because the real area of contact is less than the asperity contact area, since not all atoms in the asperity are really in contact and suggest that friction laws should be formulated in terms of the real contact area, even at the nanoscale.

To explore the effect of long-range interactions, simulations were also performed by including Van der Waals' interactions. It was found that the frictional force was again proportional to the real contact area. Surprisingly, the friction force was found to be a sublinear function of load, which would have previously been taken as confirming that elastic contact theories were valid. Simulations showed that the variation in real contact area with applied load could be successfully fitted by continuum-mechanics theories (in this case using the most general Maugis-Dugdale model).

However, the authors suggest that this model, which includes three fitting parameters, does not correctly describe the contact mechanics at the nanoscale. In this case, the sublinear dependence of friction force on load arises from the inclusion of longer-range interactions, which leads to adhesion forces that do not scale linearly with the real contact area. This results in a total load, the sum of the loads due to both short- and long-range interactions that is not proportional to the real contact area, leading to a friction force that is a sublinear function of load.

The results of these simulations suggest that the picture of the nature of the contact at the macroscopic scale, that the real contact area arises from the sum of the areas of the contacting asperities, also applies at the nanoscale.

Tribology and Lubrication Technology
June 2009, 65(6) p88

Further Reading:

[1] Tysoe, W.T. and Spencer, N.D. (2005). Searching for the Fractal Truth, Tribology and Lubrication Technology, 61(2), pp. 56.

[2] Tysoe, W.T. and Spencer, N.D. (2004). Why Does Amontons' Law Work so Well?, Tribology and Lubrication Technology, 60(8), pp. 56.

[3] Mo, Y.; Turner, K.T. and Szlufarska, I. (2009). Friction Laws at the Nanoscale, Nature, 457, pp. 1116–1119.

Split or Squeak?

A remarkably simple model goes a long way to explaining stick-slip motion

We are all familiar with stick-slip behavior, both in our professional and our everyday lives; the sound of the violin, the squeak of the horror-movie door, and the shaking during an earthquake are all manifestations of this ubiquitous phenomenon. Many studies have dealt with stick-slip and not a few attempts have been made to model it, generally focusing on the dynamics of the slider system and the slider-track interface. In a recent issue of *Tribology Letters*, Michael Varenberg and Yuri Kligerman of the Technion-IIT in Israel published an alternative, extremely simple massless (non-inertial), quasi-static (non-viscous) approach to the problem [1].

In systems in which contacting area is split among many asperities, such as under the many split protuberances in the feet of certain insects and amphibians, stick-slip has been observed to be far less likely to occur than in systems where there are few, but relatively large contacts.

In Varenberg and Kligerman's model, two situations are considered (Fig. 1). In the first case of "single contact", a block, attached to a fixed wall by a spring, is placed on a conveyer belt, and the belt set in motion. The spring force and the friction force between the block and the belt increase (stick phase) until the elastic force of the spring exceeds the static friction between the block and the belt. At this point, the block begins to slide, causing the tension in the spring to decrease until the dynamic friction value is reached so that the block sticks again, and the cycle repeats.

In the second case of "multiple contacts", the block is divided into a number of small blocks ("sub-contacts"), each being connected to the wall by a spring. This mimics the situation where contact is split between numerous asperities. This time, the stiffness and friction of each sub-contact are both fractions of the values for the single-contact case, but there exists a statistical distribution of individual values for these contacts. When any particular sub-contact enters the slip phase, as described above, the other subcontacts may still be sticking, or have already entered the slip phase.

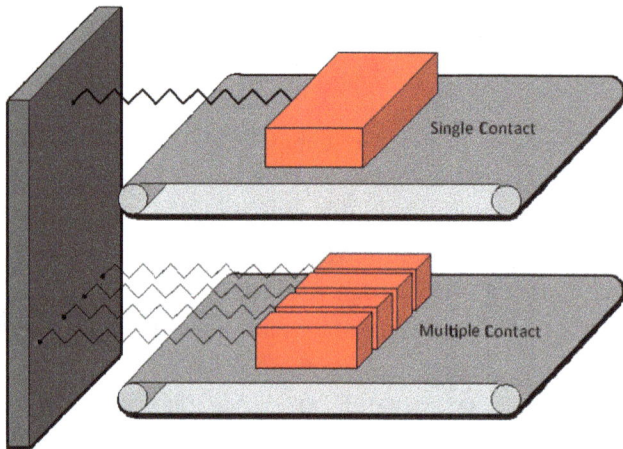

Fig. 1 Single and Multiple Contact Cases

The sum of all the friction forces is shown in Figure 2 for the cases of the single (Fig. 2(a)) and multiple (Figs. 2 (b) to (d)) contacts where here the contact is split into 2, 50, and 5000 individual subcontacts (referred to as n in the Figure). It can be seen that while the friction force displays the characteristic stick-slip behavior for the single contact, even splitting it into two contacts begins to complicate this simple behavior, due to the incoherence of the two slipping blocks. When n reaches 50 or 5000 subcontacts, the stick-slip behavior is hardly detectable anymore, even though the total friction and spring force are the same as in the single-contact case.

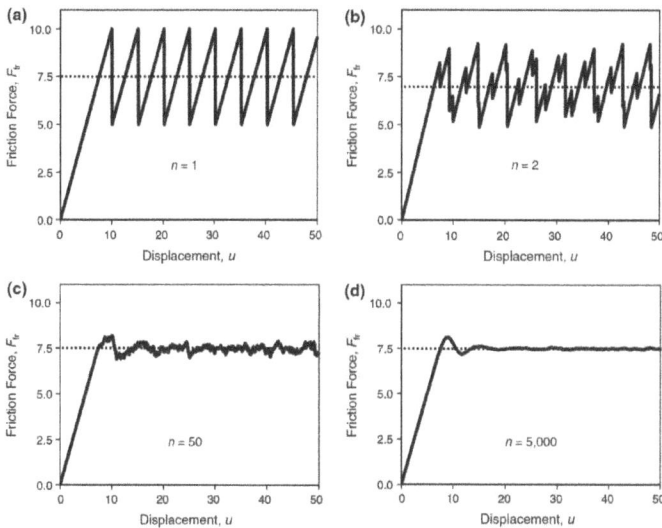

Fig. 2 Effect on friction behavior of splitting contact into *n* subcontacts. By kind permission of Springer Science+Business Media from Reference [1].

Such a simple model cannot reproduce all the properties of a real system, but it does go a long way in explaining why stick-slip behavior disappears in the case of contact splitting. It also helps us to understand the materials' influence on stick-slip and why, for instance metals, with their many characteristic asperities, are less likely to produce stick-slip than rubber-like materials, which conform better to the countersurface, and therefore form far fewer contacts.

Tribology and Lubrication Technology
April 2014, 70(4) p80

Further Reading:

[1] Kligerman, Y. and Varenberg, M. (2014). Elimination of Stick-Slip Motion in Sliding of Split or Rough Surface, Tribology Letters, 53, pp. 395–399.

Topic 8

Tribochemistry

While both of us work extensively in the area of tribology, we were both originally trained as chemists, so that the interplay between tribology and chemistry is particularly close to our hearts. While the chemistry occurring at a sliding interface is central to understanding tribological phenomena in the boundary-lubrication regime, as surface asperities come into contact during sliding, this is perhaps the most challenging area of tribology due to the difficulty of seeing what is happening in this sliding interface.

This difficulty can, to some extent be overcome by carefully analyzing the surface after sliding. This approach was illustrated in two columns that discussed the use of alcohols as lubricants. While alcohols are not usually thought of as lubricants, they do turns out to be remarkably effective for some technologically important systems. For example, 1-pentanol vapor has been found to significantly inhibit wear in microelectromechanical systems (MEMS) (*Alcohol gets you no wear*, April 2008). These miniature machines take advantage of the vast technology that has been developed by the microelectronics industry to shape and mould silicon to fabricate miniature devices that are used for accelerometers for triggering automobile air bags and in micro-mirror arrays in projectors. However, while silicon MEMS can be fabricated relatively easily, it is a terrible tribological material due to problems of excessive adhesion and wear. This article showed that the addition of 1-pentanol vapor essentially eliminated wear and the chemical processes could be elucidated by analyzing the surface using time-of-flight secondary-ion mass spectrometry (ToF-SIMS).

A similar approach was used to understand the beneficial effects of glycerol in lubricating diamond-like carbon (DLC) in *Glycerol Lubricates DLC* (August 2008). DLC generally needs no lubrication except under severe conditions, but under such conditions, glycerol proved remarkably effective. Again ToF-SIMS analyses, combined with theory in the form of molecular dynamics (MD) simulations, provided the answer.

Theory, in the form of first-principles quantum-chemical calculations, provided help in understanding the formation of "smart" films by zinc dialkydithiophosphate (ZDDP) in *Math, physics and chemistry* (August 2005). ZDDP is a widely used anti-wear additive in motor oils and the hardness of the antiwear film was found to be determined by the contact pressure. This unique behavior was explained by quantum calculations that explored the effect of pressure on the structural changes in the film.

Much tribochemical work has focused on the chemistry of oil additives. They often contain sulfur, chlorine or phosphorus, which are polluting and detrimental to the functioning of the automobile catalytic convertor. Furthermore, additives are normally dissolved in non-biodegradable oils derived from petroleum. This has prompted the search for more environmentally friendly alternatives. Water-based lubrication is clearly an attractive alternative, but its miniscule pressure coefficient of viscosity means that, when used alone, it is an awful elastohydrodynamic lubricant. The column entitled *Water: From chicken stock to base stock* (April 2004) described how mimicking nature, which very effectively uses water as a lubricant, can lead to very effective lubricants.

Carbon dioxide is usually thought of as the primary offender in global warming, so is probably not the first thing that comes to mind as a lubricant. Surprisingly, as discussed in *Lubricting with carbon dioxide* (December 2004), for steel, it lubricates rather well. Again, surface analyses provided insights into the tribochemical mechanism, in this case using X-ray photoelectron spectroscopy, which identified surface carbonates and bicarbonates as forming the lubricious coating.

The last paper in this section, *Nature's soft touch* (April 2006) highlights how difficult it sometimes is to categorize tribology into a particular area, since it touches on topics that are both biological and chemical. Soft sliding interfaces are ubiquitous in nature. Notable examples are the eye, mouth or cartilage surfaces. Soft interfaces should display low friction in water, but this is not the case for smooth silicone surfaces because they are hydrophobic, leading to strong adhesion. A chemical solution was to change the surface properties by ozone treatment to make them hydrophilic. Using chemistry to tune the nature of the surface before sliding is not the usual tribochemical approach, which generally investigates surface modification during sliding.

Alcohol Gets You no Wear

Wear is a major issue in sliding contacts in silicon-based MEMS devices. An unexpected and dramatic improvement is observed in the presence of alcohols

Microelectromechanical systems (MEMS) are becoming increasingly commonplace in applications including accelerometers that trigger the deployment of automobile airbags, the print-heads of ink-jet printers, and the micromirror arrays that lie at the heart of many data projectors. These devices are commonly fabricated by means of the silicon lithographic processes that have been brought to a high level of sophistication in the semiconductor industry. This approach offers the added advantage that a single device can incorporate both moving parts and its own control electronics.

Unfortunately, silicon is very poor material from the tribological standpoint. Despite many years of beautiful demonstrations of microturbines made of silicon, the sad reality is that there are still no commercial MEMS devices involving contacts that endure a significant amount of sliding. The main problems are either excessive adhesion and friction or unacceptable wear. These problems become increasingly significant as the size of the device is reduced, and the surface-to-volume ratio increases. While the adhesion problems can, to some extent, be alleviated by the use of fluorinated monolayer coatings, the coatings themselves wear away during the initial mechanical contacts, and require continuous replenishment. This is not a trivial undertaking — the presence of a liquid cannot be tolerated in MEMS devices, and the fluorinated compounds themselves are often solids, only generating significant vapor pressures at elevated temperatures.

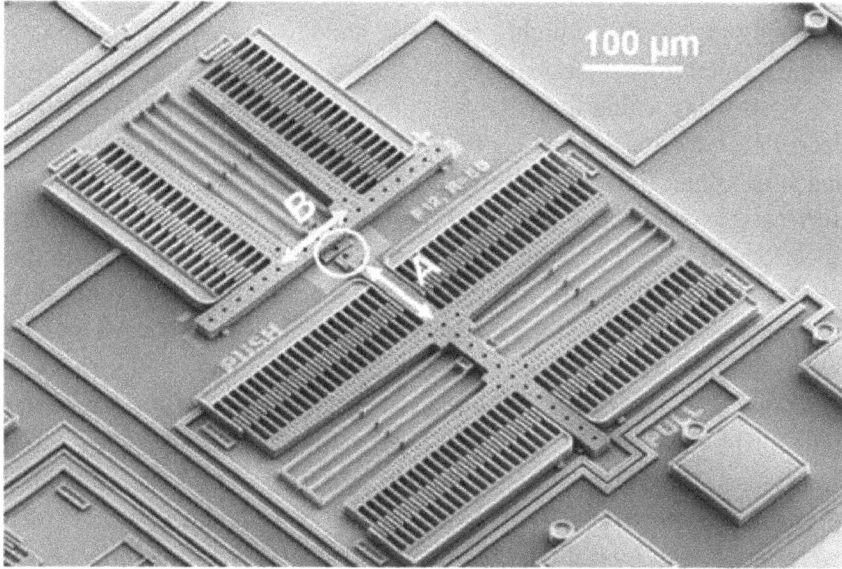

Fig. 1 SEM image of MEMS sidewall diagnostic tribometer. Sidewall contact area encircled. Loading/Unloading actuators cause the post (Labeled ''A'') to move and come into and out of contact. Push/Pull actuators cause the shuttle (Labeled ''B'') to move laterally and shear the contact. By kind permission of Springer Science+Business Media from Reference [1].

Help has arrived in the shape of Seong Kim (Pennsylvania State University), Michael Dugger (Sandia National Laboratories, New Mexico) and their team, who report their work in a recent issue of Tribology Letters [1]. Sliding cylinder-on-flat MEMS devices coated with a fluorosilane monolayer were generally found to fail after less than 10,000 cycles, at which point the actuator could no longer overcome the sliding friction. By exposing sliding MEMS devices (Fig. 1) to a continuous vapor-phase concentration of the simple alcohol, 1-pentanol, the systems were shown to continue for many millions of cycles, without failure! Moreover, the pentanol vapor seems to be able to repair failed devices: if the vapor-phase-lubricated system was stopped after a few million cycles and the alcohol replaced with nitrogen, the system failed within 10,000 additional cycles. However, upon reintroduction of the alcohol vapor, the device could be restarted, and ran for another 16 million cycles, until the operator stopped the experiment as boredom set in. This suggests that a highly effective, replenishable lubricating film is being formed from the gas phase.

What's happening in the contact? Kim, Dugger, and coworkers have recently reported some complementary surface analytical and macrotribological data in the journal Langmuir [2]. Using the very surface-sensitive ToF-SIMS (time-of-flight secondary-ion mass spectrometry) method, they observed that the contact zones in the pentanol-vapor-lubricated silicon contacts became coated with high-molecular-weight hydrocarbons during operation. The polymeric species were absent outside the contact zone, suggesting that a tribochemical reaction had taken place; similar "friction polymers" have been previously observed in oil-lubricated macroscopic systems. In linear wear tests for silicon sliding against quartz in nitrogen at contact pressures of about 100 MPa, erratic frictional behavior and high wear were observed. Upon the introduction of 1-pentanol at pressures as low as 8% of its saturation vapor pressure, this erratic behavior was not observed, friction was significantly reduced, and no wear debris could be detected.

It seems as if clever tribochemical thinking has led to something of a breakthrough in the feasibility of sliding MEMS systems and that the time is finally ripe for these potentially useful devices to deliver on their initial promise.

Tribology and Lubrication Technology
April 2008, 64(4) p64

Further Reading:

[1] Asay, D.B., Dugger, M.T. and Kim S.H. (2007). In-situ Vapor-Phase Lubrication of MEMS, Tribology Letters, 29, pp. 67–74.

[2] Asay, D.B., Dugger, M.T. and Kim S.H. (2008). Macro- to Nanoscale Wear Prevention via Molecular Adsorption, Langmuir, 24, pp. 155–159.

Glycerol Lubricates DLC

Hydrogen-free tetrahedrally coordinated carbon (ta-C) can be lubricated by liquid glycerol under boundary-lubrication conditions, to yield a friction coefficient of <0.01!

Diamond-like carbon (DLC) has found its way into many tribological applications, ranging from heart valves to fuel-injection valves. In most cases, it needs no lubrication, but increasingly there are extreme situations where a liquid lubricant needs to be applied. Conventional lubricants have generally been optimized for steel and therefore show less-than-stellar performance when used with other sliding surfaces. Enter glycerol (or glycerine): a sweet-tasting, viscous polyhydric alcohol, more closely linked with cosmetics and cake-making than with tribology.

With the exception of a body of Russian literature on the lubrication of copper-based surfaces from the 1970s, glycerol had only been investigated as a lubricant in a limited way. This situation has now been rectified by a highly productive collaboration between the group of Jean Michel Martin at the Ecole Centrale de Lyon, France, the group of Bill Goddard at Caltech and Makoto Kano, formerly from Nissan Motors Corp. Research Center in Japan [1]. Studying the lubrication properties of pure glycerol on a number of different substrates in both liquid-phase and gas-phase environments, the team was able to show that hydrogen-free tetrahedrally coordinated carbon (ta-C) could be lubricated by liquid glycerol under boundary-lubrication conditions (270 MPa and 0.3 ms^{-1}) to yield a friction coefficient of <0.01. This was more than an order of magnitude lower than could be obtained with liquid-glycerol-lubricated steel or hydrogen-containing DLC. It was also a stunningly low μ value for the boundary lubrication of anything!

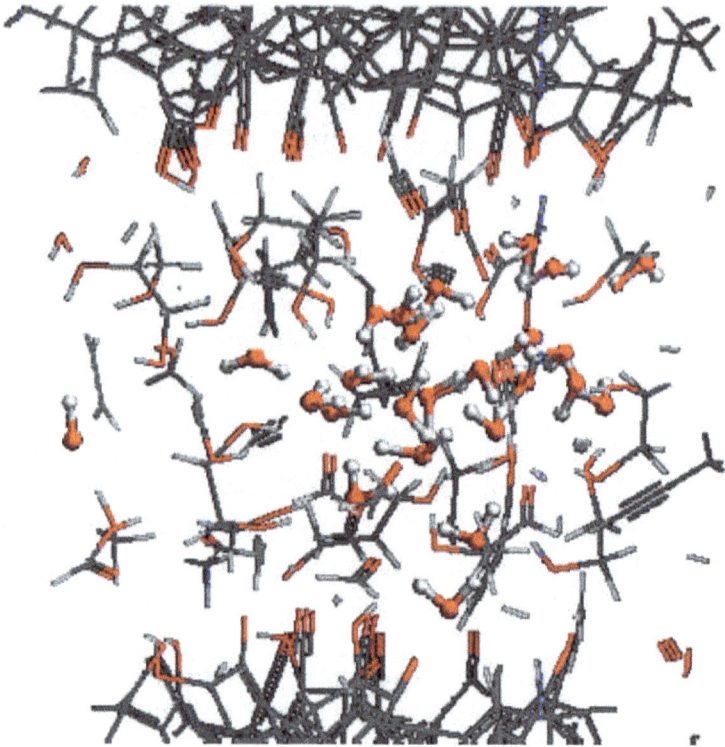

Fig. 1 Formation of water molecules by molecular dynamics simulation (shown with small circles) during lubrication of OH-terminated ta-C/ta-C in presence of initially layered glycerol molecules. Reprinted with kind permission from Reference [1]. Copyright (2008) by the American Physical Society.

A substantial battery of analytical and computational approaches was then called upon, in order to understand this surprising result. By mass spectrometrically measuring the gases released during sliding of ta-C under deuterated glycerol vapor, a substantial increase in the pressure of vapor-phase D_2O could be observed, suggesting that the glycerol was decomposing to form water under the influence of friction. Time-of flight secondary-ion mass spectroscopy showed that the wear track, following the deuterated glycerol experiment, was enriched in deuterated hydroxyl groups in comparison to the non-contact area. Also, the wear track could be far more easily wetted by water than the surrounding region. This strongly suggested that the glycerol was tribochemically reacting to hydroxylate the ta-C — a surface normally considered to be inert.

Molecular dynamics calculations of the ta-C-glycerol systems confirmed the suspicions generated by the experiments: Under shear, not only does glycerol react with the ta-C surface, but once hydroxylated, sliding ta-C surfaces exhibit extremely low friction, compared to the untreated surfaces. The model also showed that hydrogen-bonded water molecules appeared to be bridging between the surface hydroxyls.

Combining these insights, the authors proposed a mechanism that may explain the unusual and promising behavior of the system. The high pressures and temperatures at the contacting asperities lead to the generation of nascent reactive sites on the ta-C, which subsequently react with the glycerol and become hydroxylated. Glycerol itself, now in a hydrogen-bonded network, then lubricates the sliding surfaces, but finally decomposes to form organic acids and water, which then lubricates the system in turn.

Experiments by the same group have also shown that similarly record-breaking frictional values (<0.01) could be obtained with steel-steel contacts under boundary lubrication conditions (800 MPa, 3 mm s^{-1}) by adding another polyhydric alcohol, *myo*-inositol, to the glycerol at low concentrations (1 wt. %). This combination is in need of further analytical study and modeling, but potentially points the way forward to a radically new approach to boundary lubrication.

Tribology and Lubrication Technology
August 2008, 64(8) p56

Further Reading:

[1] Matta, C., Joly-Pottuz, L., De Barros Bouchet M.I., Martin, J.M., Kano M., Zhang, Q., and Goddard, W.A. (2008). Superlubricity and Tribochemistry of Polyhydric Alcohols, Physical Review B, 78, 085436.

Math, Physics, and Chemistry

Just how does ZnDTP protect surfaces? Modeling studies suggest that it may involve crosslinking of zinc phosphates under pressure

Understanding the reaction of lubricant additives at a moving interface is one of the greatest challenges in tribological research. This is partly due to uncertainties in measuring temperatures at this interface or modifications of the structure of the surface during sliding. The effect of the high pressures in the contact region, which can easily reach many Giga Pascals, has received much less attention. It is well known that pressure affects the equilibria of chemical reactions such that when there are several possible reaction products, increasing the pressure may influence the proportion of each product and therefore alter the nature and properties of the tribofilm that is formed from a particular additive.

This effect was demonstrated experimentally a few years ago by the group of Jean-Marie Georges at the Ecole Centrale in Lyon, France who suggested that the ubiquitous antiwear additive, zinc dialkyldithiophosphate (ZnDTP) forms "smart" tribofilms that exhibit a hardness that is determined by the highest contact pressure that existed during their formation under tribological stress [1]. The chemical nature of the tribofilm, and therefore its mechanical properties, depended on the contact pressure.

Professors Martin Müser and Tom Woo from the University of Western Ontario, together with graduate student Nick Mosey, addressed this question using quantum-chemical simulations, and their interesting results were recently published in *Science* [2].

They reveal that zinc phosphate crosslinks irreversibly upon compression up to around 7 GPa, to form a polyphosphate with tetra-coordinate zinc. Due to the increased connectivity of the constituent atoms caused by the high contact pressures, this material shows a much higher bulk modulus than that of the initial zinc phosphate. Up to 7 GPa, the higher the maximum pressure that the system sees, the greater the degree of cross-linking, and thus the higher the bulk modulus — in agreement with the work of Georges' group. Compression above 7 GPa shows a further, but reversible, 3-D crosslinking, as the zinc moves into a hexa-coordinate geometry. At the highest pressures simulated, in excess of 17 GPa, the bulk modulus of the film is 140 GPa — just below that of iron.

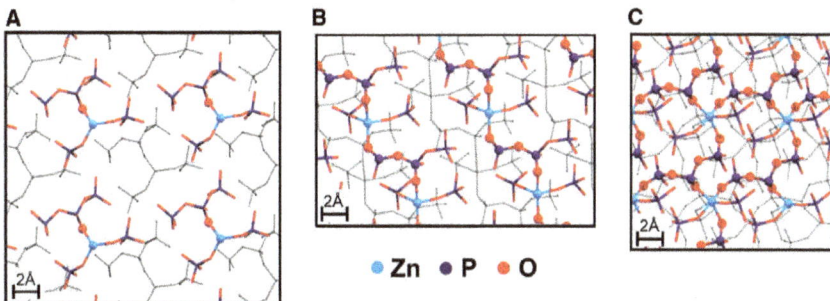

Fig. 1 Representative structures observed during the simulations of the zinc phosphate system. (**A**) The starting structure of the initial cycle. (**B**) The final structure of that cycle. (**C**) The structure of the zinc phosphate system when compressed to 17 GPa. Cross-linking in (C) occurs inall three dimensions but is only shown along the plane of the page for clarity. Color is used toindicate atoms in the uppermost layer; the atoms in the layer underneath are shown in gray. H atoms have beenomitted for clarity.From Reference [2]. Reprinted with kind permission from AAAS.

These results beautifully illustrate how contact pressures can influence tribochemistry. The consequences of this behavior are that the hard, elastic polyphosphate film formed at high pressures protects steel surfaces, but will not abrade them. The authors suggest that this would not be the case for aluminum surfaces, which, with a yield strength ≈ 7

GPa and a bulk modulus of ≈70 GPa, would not easily form the hexacoordinate zinc films, and, moreover could be damaged by any such films that formed. This could explain the much less effective performance of ZnDTP for aluminum tribosystems. The necessity for the zinc to switch between tetra- and hexa-coordination appears to be crucial, and also explains why calcium substituted for zinc in ZnDTP is ineffective as an antiwear additive.

Mosey, Müser, and Woo have shown how contact pressure can influence chemistry, but is that the end of the story? In particular, the chemistry of the total system, in particular that of the steel surface and its interactions with the reacting additive, and the potential influence of catalytic effects of the iron and the effect of relative motion between the tribological partners have not been considered. The only way to determine the importance of these effects is by experiment, and this is why calculations and experiment need to proceed hand in hand. We eagerly await the next round of experiments that test Mosey et al's seductive picture.

Tribology and Lubrication Technology
August 2005, 61(8) p56

Further Reading:

[1] Bec, S., Tonck, A., Georges, J.M., Coy, R.C., Bell, J.C. and Roper, G.W. (1999). Relationship between mechanical properties and structures of zinc dithiophosphate anti-wear films, Proc. R. Soc. London, A, 455, pp. 4181–4203.

[2] Mosey, N. J., Müser, M.H., and Woo, T.K. (2005). Molecular Mechanisms for the Functionality of Lubricant Additives, Science, 307, pp. 1612–1615.

[3] Canter, N. (2004). How does ZDDP Function? Tribology and Lubrication Technology, 61(6), pp. 20–22.

Water: From Chicken Stock to Base Stock?

Man lubricates machines mostly with oils, but nature lubricates with water. Could water be a machine lubricant of the future?

When we think of lubricants, we usually think of oils. Of course, there are plenty of oil/water emulsions used as metalworking fluids, but even there, it seems to be the oil that does most of the lubricating. Water is used to lubricate water pumps, but that is a relatively small application in the world of lubricated contacts. Given its abundance on this planet, why do we go to the trouble of basing most of our lubricants on a limited natural resource, extracted with great trouble from underneath the Kuwaiti desert or the North Slope of Alaska? The answer lies partly in a simple physical truth: water has a minuscule pressure-coefficient of viscosity, α. This essentially rules it out for elastohydrodynamic lubrication.

Nature, on the other hand, lubricates with water: Whether it is your hip joint, finger joints, knee joints, eyes, mouth or the snail's slimy trail down your driveway, water is Nature's base stock of choice. And it works! The friction coefficient in a hip joint can easily be as low as that in the best-designed oil-lubricated journal bearings. Natural lubrication manages so well with water because it relies on a combination of highly complex additives and functional surfaces, such as cartilage. The way cartilage functions is still being argued about, but it seems probable that bottle-brush structures, known as glycosaminoglycans, or GAGs, are involved. The brush hairs on the GAGs consist of chains of charged

227

sugar units, which are very heavily hydrated. It is thought that when these brushes come into contact with each other, osmotic effects inhibit their interpenetration, leading to a very low shear strength at the sliding interface and therefore low friction coefficients. Such vanishingly small friction coefficients have been observed in model, non-aqueous polymer brush systems by Jacob Klein at the Weizmann Institute in Israel [1]. Another possible mechanism is that the brushes function rather like pond weed, making the lubricant in the vicinity of the surface behave as if it had an abnormally high viscosity, thus partially compensating for the low α.

Fig. 1 The snail's slime trail is an example of natural, water-based lubrication. Photo courtesy of Taronga Zoo (http://taronga.org.au/image/snail-leaving-slime-trail), by kind permission

In an effort to see if these natural mechanisms can be translated into technological applications, one of us recently tried to lubricate steel, glass, and other materials with water containing a low concentration of a polymer (poly[*l*-lysine]-g-poly[ethylene glycol]) that is known from the bioengineering world to form water-laden brushes on many surfaces [2,3]. The results were promising, in that friction coefficients dropped by orders of magnitude, compared to using water without the additive. The denser we made the brush, the lower was the measured friction.

What about switching from oil to water? Of course, water is generally not suitable for engine lubrication or other high-temperature applications. However, there are many settings where it might be of interest, such as the food and textile industries, where contamination is a serious issue, or mining activities, where flammability is the prime concern, and water-based hydraulic fluids are already in use.

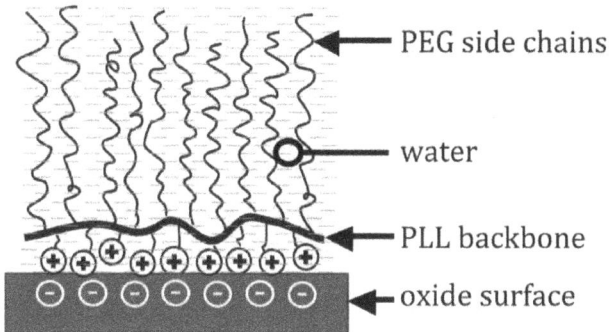

Fig. 2 Schematic diagram of poly(L-lysine)-*graft*-poly(ethylene glycol) adsorption onto a negatively charged oxide surface in aqueous solution. The water-laden brushes constitute a lubricious surface resembling lubricious surfaces in nature. Reproduced by kind permission, from Müller et al, *Macromolecules* 2005, **38**, 3861–3866.

In addition to its obvious cost, environmental and safety advantages, water is also a much better heat-transfer medium than oil. Mother Nature has developed her engineering tricks over millions of years; let's take advantage of what she has to show us!

Tribology and Lubrication Technology
April 2004, 60(4) p64

Further Reading:

[1] Klein, J., Kumacheva, E., Mahalu, D., Perahia, D. and Fetters, L.J. (1994). Reduction of Frictional Forces Between Solid Surfaces Bearing Polymer Brushes, Nature, 370, pp. 634–636.

[2] Ratoi-Salagean, M. and Spikes, H.A. (1997). Optimizing Film Formation by Oil-in-Water Emulsions, Tribology Transactions, 40, pp. 569–578.

[3] Lee, S., Müller, M., Ratoi-Salagean, M., Vörös, J., Pasche, S., De Paul, S.M., Spikes, H.A., Textor, M. and Spencer, N.D. (2003). Boundary Lubrication of Oxide Surfacesby Poly(L-lysine)-*g*-poly(ethylene glycol) (PLL-g-PEG) in Aqueous Media, Tribology Letters, 15, pp. 231–239.

Lubricating with Carbon Dioxide

Carbon dioxide appears to produce lubricious layers on steel surfaces that reduce friction, and substantially reduce wear

As environmental constraints become ever more stringent, new lubrication chemistries are being investigated to take the place of more traditional approaches involving the use of oil-additives containing metals, phosphorus, and sulfur. Metal-free additives are now commonplace in certain oil-based lubricants, and, as we have discussed in these pages before, water is being more closely examined as an alternative to oil in certain lubrication applications.

In the November 2004 issue of *Tribology Letters*, a paper from the group of Shigeyuki Mori, at Iwate University in Japan, reports on the use of carbon dioxide in a "dry" environment, to reduce friction and inhibit wear between sliding steel surfaces [1]. The investigators backfilled a vacuum chamber with various pressures of CO_2 and determined friction in an *in situ* reciprocating tester (with a 2 N applied load). Subsequently they analyzed the surface using x-ray photoelectron spectroscopy (XPS), and measured the wear on the 52100 steel disk and the 440C steel ball.

The results are intriguing, to say the least. Under optimal conditions (0.05 MPa of CO_2), the friction coefficient was reduced to $\mu=0.24$, compared to values exceeding 0.8 in air and 0.6 in vacuum. Moreover, the wear coefficients on both the ball and disk were reduced by at least three orders of magnitude!

It seems that the conditions for this behavior are quite critical. Between 0.01 and 0.05 MPa CO_2 pressures, friction is low and stable for at least an hour, and wear is reproducibly low. However, below or above this range, in a matter of minutes, the friction coefficient approaches the value obtained in vacuum,. Pretreating the surfaces at room temperature in carbon dioxide (0.1 MPa for 1 hour), followed by sliding in vacuum results in a low (μ=0.2) friction coefficient for twenty minutes or so, followed by a fairly rapid increase to the typical vacuum value ($\mu{\approx}0.6$).

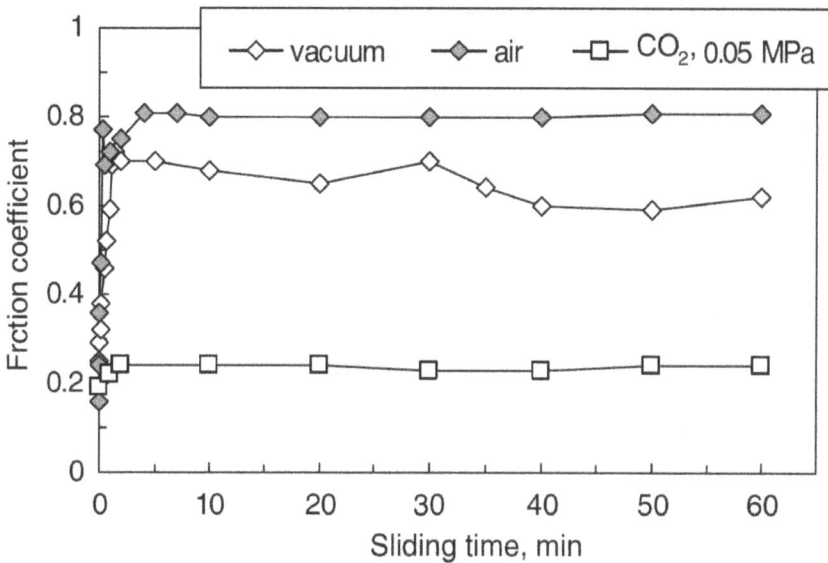

Fig. 1 Lubricating effect of carbon dioxide on a 52100 steel disk and the 440C steel ball couple. By kind permission of Springer Science+Business Media from Reference [1].

This behavior would seem to suggest that a lubricious film is formed by reaction between the steel surfaces and CO_2, even at room temperature, and that this film is worn away to some extent during sliding, requiring the film to be continuously replenished by reaction with CO_2 above a threshold pressure (0.01 MPa). The high wear rate measured at higher CO_2 pressures (>0.05 MPa) implies that chemical wear is a problem at levels of CO_2 that exceed this critical value. At these levels it would seem likely that the wear process itself leads to a higher friction coefficient.

But what is the lubricious film? XPS analysis of the 52100 disk following sliding in CO_2 shows a clear indication of a new carbon (C1s) peak that suggests the formation of iron carbonate or bicarbonate during the sliding process. Interestingly this peak is also observed on a steel surface that had been exposed to carbon dioxide without sliding, indicating that the carbonate or bicarbonate is formed even under stationary conditions at room temperature. It would seem that the carbon dioxide gas had reacted with the oxide or hydroxide present on the iron surface to produce the lubricious coating.

Currently this is at the stage of being a very interesting observation, but these results show that there are simple, alternative chemical systems that can provide useful lubrication under certain conditions. Carbon dioxide lubrication may not be the solution to all tribological problems but, as the constraints on our choice of lubricants become increasingly severe, niche solutions such as this one may be the future face of lubrication practice.

Tribology and Lubrication Technology
December 2004, 60(12) p64

Further Reading:

[1] Wu, X., Cong, P., Nanao H., Minami I. and Mori, S. (2004). Tribological Behavior of 52100 Steel in a Carbon Dioxide Atmosphere, Tribology Letters, 17, pp. 925–930.

Nature's Soft Touch

Soft elastohydrodynamic friction appears to be an important feature of natural lubrication. Nature seems to avoid hard contacts!

A couple of years ago [1], we wrote about mimicking nature by lubricating with water-laden brush systems. But water and brushes are not the only aspects of natural lubrication that set it apart from man-made approaches. Another ubiquitous characteristic of natural tribosystems is that they involve *soft* surfaces, such as the eye, the inside of the mouth, the surface of cartilage or the "foot" of the snail. In addition to the water-laden brush systems, the low elastic modulus of the surfaces involved seems to be a key contributor to the very low coefficients of friction (μ) observed in nature. Water-based lubrication between soft surfaces often involves the *soft-elastohydrodynamic* ("soft-EHL") or *isoviscous-elastic* lubrication mechanism. An important characteristic of this regime is the low contact pressure, which is sufficient to distort at least one of the surfaces, but insufficient to affect the lubricant's viscosity.

Recently, research carried out in one of our laboratories [2] has shown that *both* the surface chemistry *and* the soft-EHL mechanism are responsible for the very low μ values encountered in nature. Duncan Dowson and Bernard Hamrock, at the University of Leeds, UK, had already shown [3] that one can calculate the film thickness between soft-EHL-lubricated surfaces, and predicted that the lubricant film thickness between two pieces of silicone rubber sliding in water at 1 cm/sec under a 1 N load, is around 15 nm. In the case of very smooth silicone surfaces, with a root mean square roughness on the order of 2 nm, this should lead to full fluid film lubrication, and thus a low μ. Actual measurements, however, show a very high friction coefficient of around 2! Silicone, however, is fundamentally different from the soft surfaces found in nature in terms of its *hydrophobicity*. This means that the surfaces will adhere strongly to each other in water, leading to very high friction. The easiest solution to this problem is to oxidize the silicone surfaces in an oxygen plasma cleaner prior to the measurement (making them more *hydrophilic* and therefore less adhesive to each other in water*)*, which changes the lubrication mechanism such that fluid-film lubrication is indeed achieved, lowering the friction to around 0.04 at 10 cm/sec. As the sliding speed is lowered, one would expect that the fluid film would decrease in thickness so that the asperities at the surface would, at some point, start to interact, leading to an increase in μ. This is exactly what happens, and by 0.01 cm/sec, μ has increased to 0.1 (the boundary-lubrication regime).

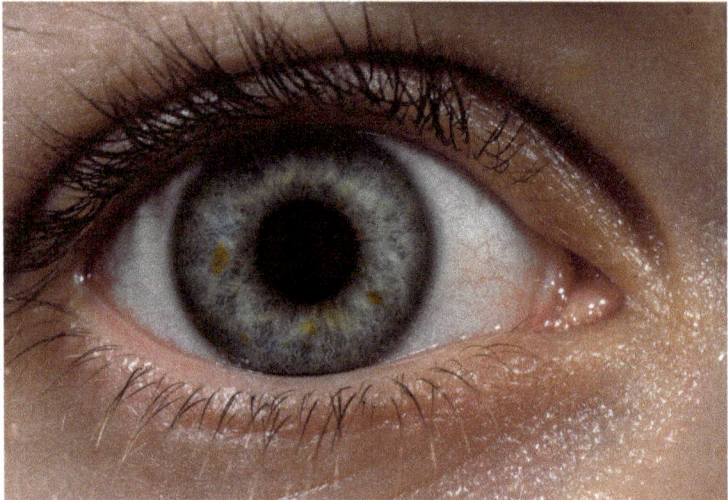

Fig. 1 Example of a natural, soft lubricated surface: Human eyeby ROTFLOLEB — Licensed under Creative Commons Attribution-Share Alike 3.0 via Wikimedia Commons

The most interesting effect is seen when one takes one more step towards imitating nature and covers the plasma-oxidized surfaces with dense brushes of poly(ethylene glycol), which retain a layer of water bound to the surface. In this case, the friction measured at high speeds is the same as for the brush-free case, but now at low speeds (tested minimum speed 0.01 cm/sec), the friction does not increase. In other words, it is as if the brushes prevent the transition to boundary lubrication at low speeds, maintaining frictional coefficients typical of fluid-film lubrication! It is thought that this occurs due to the high, entropically induced forces acting against the compression or interpenetration of polymer brushes.

Clearly the combination of brushes and soft surfaces is important in nature, where the brushes are composed of polysaccharides, rather than poly(ethylene glycol). The challenge now is whether any useful mechanical systems could be redesigned with soft, brush-covered contacts, to take advantage of these new, biomimetic lubrication effects!

Tribology and Lubrication Technology
April 2006, 62(4) p56

Further Reading:

[1] Tysoe, W.T. and Spencer, N.D. (2004). Water: From chicken stock to base stock, Tribology & Lubrication Technology, 60(4), p64.

[2] Lee, S. and Spencer, N.D. (2005). Aqueous Lubrication of Polymers: Influence of Surface Modification, Tribology International, 38, pp. 922–930.

[3] Hamrock, B.J. and Dowson, D. (1979). Minimum film thickness in elliptical contacts for different regimes of fluid-film lubrication, Proc. 5th Leeds-Lyon Symp.Tribol., MEP, pp. 22–27.

Topic 9

Weird and Wonderful Effects in Tribology

A large proportion of tribologists work in the more traditional areas of oil lubrication of machine elements or wear of steel components. Nowadays, many others are to be found among the ranks of those engaged in research in articular implants or hard-disk systems, as well as fundamental aspects of tribology. In addition to these important, and now mainstream areas, however, we have endeavored, over the last decade, to bring less obvious tribological topics to our readers' attention.

A fascinating area of tribology is that involving ice and snow, and it is one that has capitivated tribological scientists since Bowden's pioneering studies on skiing in the 1950s. It is astonishing that new insights continue to be gained today, and our two articles in this area — on skiing (*Tribology's Olympic Research*, April, 2010) and on curling (*Why do Curling Stones Curl?*, August, 2013) — discuss groundbreaking research in the tribology of winter sports, and show that there is still much to be done to fully understand the phenomena at work when sliding on solid surfaces near their melting point.

Tribology is increasingly important in a number of large-scale technologies that we do not normally associate with our field. Two of these, rarely addressed in the same context, are cosmetics (*The Tribology of Personal Care*, December, 2009) and cement (*Tribology Influences Rheology Influences Tribology...*, December, 2013). In the former column, the emphasis is on the tribology of sensory perception, while in the latter, the all-important rheology of pumping cement slurries is dealt with by considering the tribology between individual particles.

Electron emission has been known for some time to occur during tribological contact, and in our piece entitled *Where are all the electrons from?*, October, 2014, we described observations of distinct spatial-distribution patterns of electron emission during a pin-on-disk experiment. These include emission from the inlet of the contact, straight lines of emission across the sample (suggesting that cracks play a role in the process), and a persistent emission that was probably due to surface-charging effects.

Two surprising tribological observations were described in our pieces on x-ray emission upon peeling Scotch tape (*X-rays by Triboluminescence*, April, 2009) and friction forces occurring out of contact (*Friction at a Distance*, February, 2012). While these effects may not influence our everyday lives, or change tribological practice overnight, they show the ubiquity of tribological phenomena, and the all-important role of fundamental physics in explaining them.

Finally, our column on one of the many seasonal tribological irritations involved in travel (*Leaves on the Line*, December, 2006) describes a beautiful study where engineering, chemistry and botany come together to explain why leaves lead to the spinning of steel wheels on rails.

Why Do Curling Stones Curl?

Tribologists in Sweden debunk old theories — and come up with a new one that fits all the facts!

Curling is an Olympic sport, played on ice, which is enormously popular in northern countries, such as Canada, Scotland, and Sweden. It involves two teams who slide a series of 16 stones along a 28 m stretch of ice, aiming to get them as close to the target as possible. Once the stone is released at one end of the ice sheet, the trajectory is governed purely by sliding friction. The subtlety of the game is that by rotating the stone upon release (producing 1-3 rotations over the entire 28 m), the stones can adopt a curled trajectory, thereby avoiding "guarding" stones that would otherwise prevent the moving stone from accessing the target (see Figure 1).

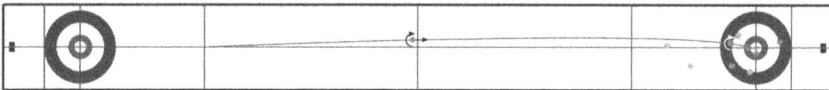

Fig. 1 The trajectory of a curling stone. By kind permission of Springer Science+Business Media from Reference [1].

Harald Nyberg, of the Ångström Tribomaterials Group, at the University of Uppsala, Sweden, together with colleagues Sture Hogmark, Staffan Jacobson, and Sara Alfredson, have recently published two landmark papers on this inherently tribological sport. One [1] of these discounts old theories, while the other [2] proposes a novel theory that appears to account for all observations to date. It is important to know that the bottom of the stone is hollowed out, such that the contact with the ice is made with an annulus that is around 6 mm wide, and 120 mm in diameter. The surface of this annulus is intentionally roughened during stone preparation, and the degree of roughness is known to influence both friction and degree of curl, which can be as much as 1.2 m lateral deviation from a straight trajectory. The preparation of the ice is also important: the surface is pre-sprinkled with water, resulting in the formation of bumps or "pebbles" (between 0.2 mm and 5 mm wide) whose tops are flattened by a special tool prior to the game.

In their first paper [1], the authors point out that most theories to date have been based on the presence of a friction asymmetry between the leading and trailing halves (as seen in the direction of the motion) of the annulus. This may seem reasonable at first, but it should be borne in mind that if, for example, an inverted drinking glass is slid along a table with a rotating component to its motion, the resulting curl is exactly *opposite* to that observed for a curling stone. In other words, there must be another, much greater effect at work. Previous studies have explored the influence of meltwater layers on the friction, as they become dragged around with the rotating annulus, but there remain many open questions about exactly how this may lead to a curl.

Nyberg et al decided to construct a numerical model that checks whether a friction asymmetry, irrespective of its origin, could ever lead to the curling effects observed in practice. They found that even in the most extreme case imaginable, where *all* friction force was acting on the trailing half of the stone annulus, the observed curls could not be reproduced. Furthermore, their model showed a pronounced dependency of curl on rotational speed, which curlers know to be a non-significant parameter.

In the second paper [2], the authors invoke a completely new mechanism that takes into account the roughness of the stone, since they made the experimental observation that polished stones do not curl at all. They also found, by electron microscopic analysis of ice replicas, that the surfaces of the ice pebbles are scratched following the passage of a curling stone. Furthermore, they noticed that non-rotating stones could be given a curl by prescratching the ice in a particular direction. This latter effect appeared to diminish with repeated experiments, suggesting that the scratches were being worn away.

The mechanism that leaped out from these observations was as novel as it was all-encompassing: The roughness asperities on the leading edge of the stone were scratching the ice-pebble surfaces. Then, as the stone rotated, the asperities on the trailing edge encountered the scratches formed in the ice, and were guided by them, thereby leading to a curl in the correct direction. The forces encountered by the asperities contacting the scratches could be calculated by simple mechanics, and were shown to be indeed of sufficient magnitude to induce the curl. The mystery of how curling stones curl appears to have been laid to rest!

Tribology and Lubrication Technology
August 2013, 69(8) p72

Further Reading:

[1] Nyberg, H., Hogmark, S. and S. Jacobson (2013). Calculated trajectories of curling stones sliding under asymmetrical friction — validation of published models, Tribology Letters, 50, pp. 379–385.

[2] Nyberg, H., Alfredsson, S., Hogmark, S. and Jacobson, S. (2013). The asymmetrical friction mechanism that puts the curl in the curling stone, Wear, 301, pp. 583–589.

Tribology's Olympic Research

A recent study sheds new light on the interaction of materials and roughness effects on ski friction

With the Winter Olympics behind us, it may be time to reflect on the extent to which tribology has contributed to the sports we have enjoyed watching. Clearly the major reason for playing sports on ice and snow at all (rather than staying at home drinking hot cocoa in front of the fire) is that frozen water offers the possibility for human beings to move across the land at very high speeds. Skiing, which has its origins as far back as the seventh millennium BC, was introduced as an Olympic sport in the 1924 Winter Games in Chamonix, and remains a very popular winter pastime in many parts of the world. While wooden skis were the norm up until the middle of the 20th century, polymers began to be used as a ski base in the 1950s, and ultrahigh molecular weight polyethylene (UHMWPE) has remained the standard surface for snow contact, both for amateurs and professionals.

F.P. Bowden, of the Cavendish Laboratory at Cambridge University, was the first tribologist to examine the processes taking place at the ski-snow interface, and his early observations, beginning in the 1930s, are still cited heavily in the skiing-tribology literature today [1]. His theory that skis are actually lubricated by melt-water, generated by friction at the ski's leading edge, is still regarded as the most likely cause of the very low friction between ski and snow. He also tested the effect of different ski-base materials on ski speed, finding that among those tested, the fastest material was PTFE — the most hydrophobic material tested.

242

The hydrophobicity is widely believed to inhibit capillary drag between the ski and the meltwater-covered surface. Although attempts have been made to use this material since Bowden's work, PTFE suffers from poor abrasion resistance, and has therefore not supplanted UHMWPE.

It is common practice among skiers to prepare their ski surfaces by roughening them. A number of studies have examined the influence of roughness on ski friction, but until recently, no studies had dealt with the interaction of materials and roughness effects. This situation was recently remedied by a fascinating study carried out by Jan Giesbrecht, Theo Tervoort, and Paul Smith at ETH Zurich, Switzerland [2]. Using a custom-designed test ski on a real, freshly prepared Nordic ski track, they compared a series of different polymers with low roughness and different water contact angles and confirmed Bowden's observation that the more hydrophobic the surface, the lower the friction and the faster the ski.

Fig. 1 Freestyle Skiing. By 极博双板滑雪俱乐部 (Uploaded to Flickr as jfpds regular) [CC-BY-SA-2.0 (http://creativecommons.org/licenses/by-sa/2.0)], via Wikimedia Commons.

The next step was to introduce roughness, and this is where the results took an unexpected turn. Under the conditions investigated, at roughness values below 0.2 μm (R_a), the behavior was exactly as described above, but in the roughness range 0.2-1 μm, the speed increased significantly, and was completely independent of the material! Above 1 μm roughness, the speed reduces again, and becomes dependent on the directionality of the roughness. These findings suggest that capillary suction is the main contributor to friction for smooth ski bases, while plastic deformation of the snow is the main dissipation mechanism at roughness >1μm. The "sweet spot" lies between these two regimes, although it does depend on temperature, since this also determines the amount of meltwater present on the snow surface. The consequence is that different degrees of roughness are appropriate for different temperatures of snow — a fact that was empirically determined by professional skiers some time ago.

This research has opened new opportunities for ski design, since it shows that designed roughness can be used as a tool to overcome hydrophilicity concerns for polymers that might otherwise have useful properties for skis, such as high abrasion resistance.

Tribology and Lubrication Technology
April 2010, 66(4) p56

Further Reading:

[1] Bowden, F. P. and Hughes, T. P. (1939). The Mechanism of Sliding on Ice and Snow, Proc. R. Soc. London, 172, pp. 280–298.

[2] Tervoort, T., Giesbrecht, J., and Smith, P. (2010). Polymers on Snow: Towards Skiing Faster, J. Poly.Sci. B, 48, pp. 1543–1551.

The Tribology of Personal Care

Researchers are improving cosmetic products with the help of novel tribometer measurements

While many in our community concern themselves with the sliding contact of metallic components, their lubrication with oil, and all the associated sealing, condition-monitoring, and additive technologies, we should not forget that tribological issues go far beyond this, into the most personal aspects of our everyday lives. Tribological problems in the personal-care industry abound, whether they be in the design of hair-care products (which probably also involves some condition monitoring…), the development of nail polish, where adhesion and wear phenomena are important issues, or the formulation of skin creams, where the lubrication behavior of the cream during application can determine the degree of consumer acceptance.

In a recent paper in Tribology Letters, Ken Nakano and his coworkers, from Yokohama National University and the Kao Corporation in Japan, have looked into the tribological phenomena involved in the application of cosmetic foundation [1]. A mainstay of the industry has been the use of sensory panels, who report on the tactile sensations involved during the application of various skin products. While this information is useful, it tends to be subjective and vague and involves a large number of panel participants. Nakano's work represents a step towards more objective prediction of sensory response. It involved the correlation of sensory-panel data with measured tribological quantities from a specially designed tribometer, by means of multiple-regression analysis.

The sensory panel was asked to rub a number of formulations resembling cosmetic foundation into the inner surface of the wrist with a forefinger, and to report the sensation and comfort level according to smoothness, silkiness, velvety feeling, softness, and skin-adhesion ability. The formulations consisted of a mixture of mica particles (10 × 0.2 μm platelets), acrylic spheres (5μm dia.), and silicone oils in different ratios.

Fig. 1 Cosmetics in the fine arts: *Femme à sa toilette*, Henri de Toulouse-Lautrec (1889).

The tribometer involved sliding a silicone rubber sphere against a silicone rubber plate (both with moduli of 0.5 MPa), both sides being supported by double-cantilever springs, lasers being used to monitor displacement in both x and z directions. A number of different experiments were carried out, including a dry-sliding control, a fully lubricated test, with the sample on the plate, a starved lubrication experiment, with the sample on the sphere, and another fully lubricated experiment with grooves on the plate, to simulate a fingerprint structure. The quantities measured included static and kinetic friction, but also dynamic-response parameters, such as the time required to reach asymptotic values of friction, and the amplitude/degree of damping of the vibrations due to natural system resonance and the oscillations caused by sliding over the simulated fingerprint structure. A minimum value of kinetic friction was found for a composition capable of forming a layered structure of spheres and platelets.

By means of multiple-regression analysis, the sensory-panel data were correlated with the tribometer measurements. A reasonable degree of correlation between the two datasets could be obtained, despite a simple linear correlation being used to investigate the very non-linear behavior of our senses. Some interesting conclusions could be drawn: Skin-adhesion ability, for example, was strongly correlated to the degree of damping of the natural system frequency. A surprise was that comfort was not simply related to static and kinetic friction. Greater comfort was in certain cases reported for higher static friction and also greater vibrational amplitudes. Next steps will involve the use of non-linear approaches, such as neural networks, to better model the sensory response.

It is clear to all of us that tribology is far from being a straightforward field. The tribology of sensory phenomena represents a significant added level of complexity, with many challenges and opportunities, in terms of greater understanding of our senses, and improved products for the personal-care industry.

Tribology and Lubrication Technology
August 2009, 65(12) p64

Further Reading:

[1] Horiuchi, K., Kashimoto, A., Tsuchiya, R., Yokoyama, M., and Nakano, K. (2009). Relationship Between Tactile Sensation and Friction Signals in Cosmetic Foundation, Tribology Letters, 36, pp.113–123.

Tribology Influences Rheology
Influences Tribology...

Applying a new tribological model to understanding paste rheology not only fits the experimental facts, but it points the way to better-flowing cement slurries.

The fact that the rheological properties of lubricants have a significant impact on tribological behavior is well known. Tribology influencing rheology is not so obvious, but it lies at the heart of a new theory of flow in dense particle suspensions, just published by a research group involving the ETH Zurich, Switzerland, the Ecole Centrale de Lyon, France, and the French cement manufacturer Lafarge [1].

Paste rheology is important in many processes, but it is critical in the construction industry, where dense cement slurries often need to be pumped long distances into inaccessible places, or to significant heights (e.g. to the top of the recently completed Burj Khalifa in Dubai, 829.8 m). Above a certain particle density, pastes under flow show an increase in viscosity with increasing shear rate. This is known as shear thickening, and, at high pumping rates, leads to a much higher energy consumption than would be expected for a simple fluid. As if this were not bad enough, under certain circumstances the viscosity can increase to infinity. This phenomenon is called discontinuous shear thickening (DST) and can be thought of as a jamming process that completely prevents the paste from flowing. This quickly leads to equipment failure, frustrated builders, hysterical architects, and angry investors.

To get to the bottom of this behavior, and to try to find a solution to the problem, the Swiss and French researchers looked at flowing cement slurries from a tribological standpoint. The rheological behavior of dense suspensions in a pipe is dominated by interactions between the solid particles in the slurry, which either glide around or scrape over each other, the latter being more likely to happen at higher shear rates. If the particle concentration and shear rate are sufficiently high, this ultimately leads to DST. Thinking in terms of the Stribeck curve, the low-shear-rate behavior can be regarded as hydrodynamic lubrication, while the high-shear-rate behavior is more reminiscent of the boundary regime (Fig. 1).

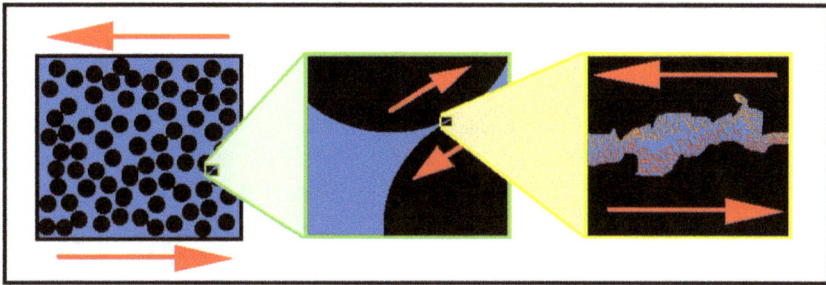

Fig. 1 The rheology of a concentrated particle suspension (left) can be considered in terms of the tribology of a set of particles rubbing against each other in a fluid (middle), applying the usual considerations of rough surfaces and boundary lubrication (right). By kind permission of Nicolas Fernandez, ETH Zurich.

This theory was put to the test both experimentally and with a numerical model. Both showed that lowering the friction coefficient between particles in a slurry allowed flow to occur with higher solids loadings without the dreaded DST occurring. This has important practical consequences, since it means that the cement manufacturer can add small quantities of surface-adsorbing polymers to form a so-called polymer brush on the surface of the particles. These layers significantly reduce particle-particle friction, minimize the likelihood of DST and therefore allow denser cement pastes to be transported through pipes at higher speeds and over longer distances. It appears that tribological research not only contributes to improving the efficiency of automobile engines and machines, but also in the construction industry.

Tribology and Lubrication Technology
December 2013, 69(12) p96

Further Reading:

[1] Fernandez, N., Mani, R., Rinaldi, D., Kadau, D., Mosquet, M., Lombois-Burger, H., Cayer-Barrioz, J., Herrmann, H.J., Spencer, N.D. and Isa, L. (2013). Microscopic mechanism for the shear-thickening of non-Brownian suspensions, Phys. Rev. Lett., 111, 10830.

X-rays by Triboluminescence

A recent discovery shows that more than just light can be produced by rubbing

The phenomenon of triboluminescence, the emission of light during rubbing, is well known and was first recorded by Francis Bacon in 1620, where he noted that "It is well known that all sugar, whether candied or plain, if it be hard, will sparkle when broken or scraped in the dark". Also, the emission of other particles such as electrons to produce a "triboplasma" at sliding interfaces has been well documented. One might expect that this phenomenon would be related to the strength of the chemical bonds that are broken. However, it was found as early as 1939 that an adhesive tape affixed to a surface, which is thought to be held together by rather weak Van der Waals' forces, provided, upon peeling apart, an example of a triboluminescent system and generated light emission that could also be detected by the naked eye.

Even more surprising was the recent discovery by Professor Seth Putterman and his group at the Department of Physics and Astronomy at the University of California, Los Angeles that X-rays could be generated as well. In order to investigate this phenomenon, they constructed an apparatus that enabled them to peel a roll of commercial Scotch® tape at a controlled velocity (Fig. 1) [1]. Since they found that the effect was quenched by air, the experiment was performed under a moderate vacuum of about 10^{-3} Torr (about a millionth of atmospheric pressure), while detecting the resulting emitted X-rays using an efficient, high-speed X-ray detector.

They observed the emission of X-ray pulses lasting only about a billionth of a second that correlated with slips in the force that was required to peel the Scotch tape. While the X-ray photons had energies of about 20 keV, the pulses could contain more than 100,000 photons, resulting in a total energy per pulse of GeVs. Based on the length of the X-ray pulses, the authors suggested that the emission occurred from regions at the point of peeling that were much smaller than a millimeter.

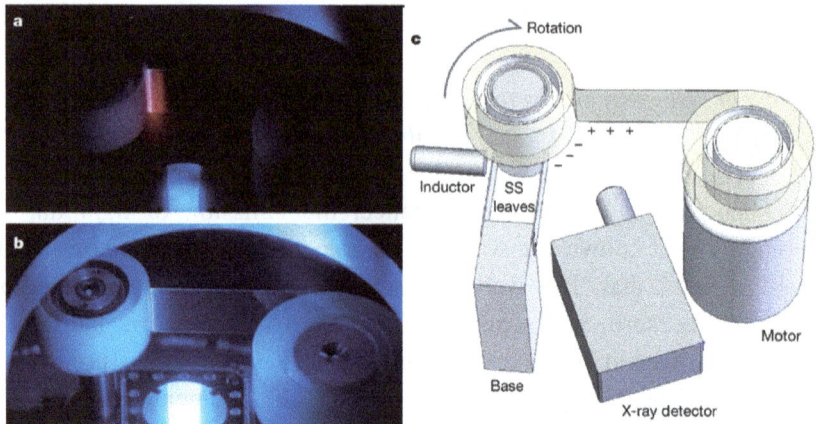

Fig. 1 **a,** Photograph of the simultaneous emission of triboluminescence (red line) and scintillations of a phosphor screen sensitive to electron impacts with energies in excess of 500 eV (under neon at a pressure of 150 mTorr). **b,** Photograph of the apparatus (under a pressure of 10^{-3} Torr) illuminated entirely by scintillations. **c,** Diagram of the apparatus used to measure peeling force. Reprinted by kind permission from Macmillan Publishers Ltd: Nature (Reference [1]), copyright (2008).

The authors observe that, during peeling, the adhesive becomes positively charged, while the underlying polyethylene roll becomes negatively charged, so that large electric fields are built up and become sufficiently strong to trigger discharges. Under the reduced pressures of the experiment, the electrons in the discharge are accelerated by the electric field between the roll and the adhesive to sufficiently high energies that they emit X-rays as they strike the positive side of the tape.

While a sensitive X-ray detector was used to monitor the emitted X-rays and to measure their energy distribution and duration, the emitted X-ray flux is quite high. A very striking demonstration of this significant intensity was made by taking X-ray images of one of the author's fingers, where they were able to collect quite clear images of the bone in the finger with only a 20 s exposure while a tape was being peeled at 10 cm/s. The process by which the high-energy photons (X-rays) are emitted is still not completely understood but the authors suggest that these effects might lead to insights into the fundamental aspects of tribology. In the meantime, fundamental research into triboluminescent X-ray emission might help to limit rising health-care costs!

Tribology and Lubrication Technology
October 2009, 65(10) p72

Further Reading:

[1] Camara, C.G., Escobar, J.V., Hird, J.R. and Putterman, S.J. (2008). Correlation Between Nanosecond X-ray Flashes and Stick-Slip Friction in Peeling Tape, Nature, 455, pp. 1089–1092.

Friction at a Distance

Superconducting materials enable probing of out-of-contact friction mechanisms

Since tribology is defined as "the study of contacting bodies in relative motion", we generally think of friction as occurring only when two surfaces touch. However, two closely spaced objects still interact, even if they are not in contact. This implies that energy can be dissipated when two closely spaced objects move relative to one another, producing a friction force, and such "non-contact" friction has indeed been observed.

There are two possible ways in which energy could be dissipated in this non-contact regime. The first is by electronic friction. When two different materials come close to each other, charge is transferred from one to the other. When they move relative to each other, this charge can fluctuate to produce a fluctuating current. Just as an electric current passing through a resistive wire loses energy to produce heat, when two surfaces move relative to each other, the fluctuating current produced also dissipates energy.

Second, the atoms in the two materials can also interact at a distance. When the two closely spaced surfaces move relative to each other, atomic vibrations are excited, and these motions can also dissipate energy. Since such vibrations in solids are known as "phonons", this is known as phononic friction.

Although non-contact friction has been detected, which of the two described mechanisms dominate the friction is not well understood. In principle, this could be decided by measuring the properties of many different materials and by correlating their friction behavior with their electronic and vibrational properties.

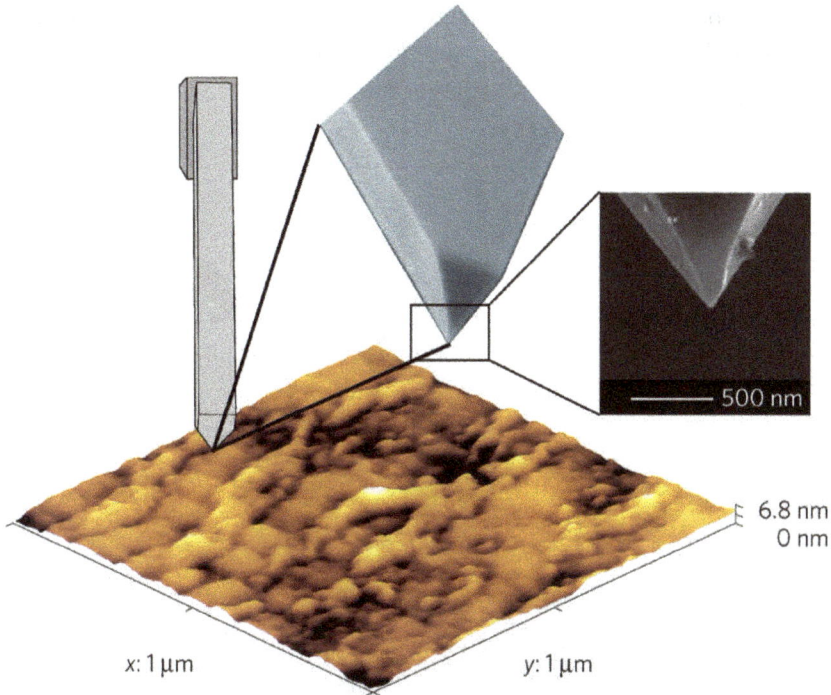

Fig. 1 AFM topography image of the Nb film studied in the experiment. Reprinted by kind permission from Macmillan Publishers Ltd: Nature Materials (Reference [1]), copyright (2011).

Professor Ernst Meyer and his group at the University of Basel in Switzerland have used an alternative and more elegant approach. They measured non-contact friction using a niobium sample (Fig. 1). Niobium is a superconductor, meaning that above the superconducting transition temperature, T_C (which is at 9.2 degrees Kelvin), it behaves like a resistive metal, while below this temperature, the resistance drops to zero. Thus, if friction is dominated by electronic effects, it should decrease dramatically as the temperature drops below 9.2 K.

They measured the friction using a silicon tip that was placed 0.5 nanometers from a very flat niobium sample. In order to maintain this small distance accurately, they measured friction by laterally exciting the tip like a pendulum at its resonant frequency (of 5.3 kHz). They then measured the time for the tip amplitude to decay when the driving frequency was switched off; the larger the friction force, the more rapidly the tip amplitude would decay.

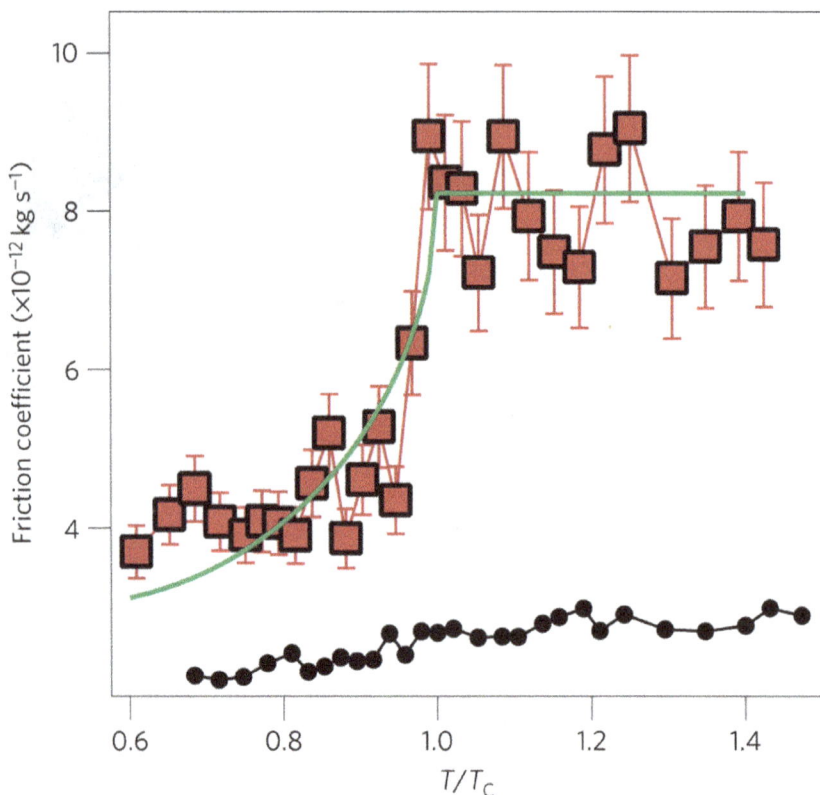

Fig. 2 Temperature variation of the friction coefficient across the critical point $T_c = 9.2$ K of Nb. Reprinted by kind permission from Macmillan Publishers Ltd: Nature Materials (Reference [1]), copyright (2011).

By carefully measuring the friction force in small temperature steps below and above T_C, they found that the friction was much lower (by a factor of about four) below T_C than above, suggesting that electronic friction dominates (Fig. 2).

They also measured the friction as a function of distance d from the surface for a temperature below (5.8 K) and above (13 K) T_C. Above T_C, where metallic behavior was found, the friction force is proportional to $1/d$, exactly what has been predicted for electronic friction. Below T_C, the friction force was found to decay as $\sim 1/d^4$, consistent with theories for phononic dissipation.

These non-contact friction forces are many orders of magnitude lower than contact friction so that engineers need not concern themselves with designing lubricants to mitigate their effects. However, they may become important in micro and nano-electromechanical systems.

Tribology and Lubrication Technology
February 2012, 68(2) p72

Further Reading:

[1] Kisiel, M., Gnecco, E., Gysin, U., Marot, L., Rast, S., and Meyer, E. (2011). Suppression of Electronic Friction on Nb Films in the Superconducting State, Nature Materials, 10, pp. 119–122.

Leaves on the Line

"Leaves on the line" constitute a major tribological problem for railways in the fall. The mechanisms behind the phenomenon involve an unfortunate combination of cellulose, pectin and iron.

In the country from which we both hail, an annual excuse for delayed trains has always been "Leaves on the line" (LOL). This is a problem that is not limited to the UK, however, and many railways throughout the world have to deal with service delays in the fall, caused by leaf-residue films leading to a loss of wheel-track adhesion. The films, which are hard, glazed, black and extremely difficult to remove from the rails, have been described as "Teflon-like" and can seriously influence braking distances. Approaches to removing the films have ranged from low-tech (sand) to high-tech (laser ablation), and, of course, the last-resort option of tree removal. Only the latter has proven to be totally effective, and thus the problem persists. In Sweden alone, the LOL-incurred delays have been estimated to cost over \$13 million annually, but surprisingly, remarkably little fundamental research has been carried out on the problem. What is known is that the black film contains iron, iron oxide, cellulose, water, and oil. The friction coefficient between wheel and rail under normal, dry conditions is between 0.45 and 0.65, dropping to $\mu < 0.3$ under wet conditions. The film further reduces μ to around 0.1 under dry conditions, which can drop to as low as 0.02 in the presence of dew, coinciding, of course, with the morning commute to work.

Dr Philippa Cann, of Imperial College, London, and a daily rail commuter, has now investigated LOL from a tribochemical standpoint, taking real (sycamore) leaves, chopping them into small pieces and soaking them for up to two weeks in water, the supernatant, which became quite viscous, being preserved for later tests [1]. The degraded leaves were placed, with water droplets (simulated dew), between steel surfaces in a sliding/rolling ball-on-disk tribometer. At a contact pressure of around 1 GPa, speeds up to 2 mph, and a slide/roll ratio of up to 50%, she was able to operate within the conditions encountered in a real wheel-rail system. A black residue formed rapidly, and its tribological properties were explored before subjecting it to analysis with infrared reflection-absorption spectromicroscopy.

Fig. 1 An autumn leaf, ready to cause rail havoc

The tribometer tests confirmed the field tests and showed the leaf residue to reduce µ to as low as 0.01. Interestingly, the supernatant also functioned as an effective lubricant at higher speeds (presumably due to its viscosity), but was less lubricious than the residue at low speeds. The supernatant was also found to produce the characteristic black coloration upon reaction with the steel surfaces.

The infrared tests unambiguously showed the presence of pectin in both the black deposits and in the supernatant. Pectin is a heterosaccharide found in the cell walls of plants, and is known to form gels in the presence of metal ions (such as the iron from the steel rail). The deposits were also found to contain cellulose. Pectin is known to flocculate cellulose in the presence of iron ions, and this would seem to be the mechanism at work in LOL as well. The combination of cellulose and pectin gels appears to be particularly effective at lubricating the wheel-rail contact, since the cellulose provides a solid-state boundary lubrication function, while the gels provide viscosity for elastohydrodynamic lubrication at higher speeds.

Now that the composition of the LOL films has been revealed, it will be much easier to propose strategies for their removal, perhaps using (bio)chemical reactants.

Tribology and Lubrication Technology
December 2006, 62(12) p56

Further Reading:

[1] Cann, P.M. (2006). The "leaves on the line" problem - a study of leaf residue film formation and lubricity under laboratory test conditions, Tribology Letters, 24, pp. 151–158.

Where are all the Electrons from?

*Imaging the distribution of triboemitted electrons in vacuum enables the
sources of the electron emission to be identified*

A previous column discussed the intriguing observation that X-rays were
emitted when Scotch tape was rapidly peeled [1]. However, elementary
particles such as electrons have also been detected when surfaces are
rubbed. One model for electron emission suggests that they form due to
surface fracture, during which a charge imbalance between the faces of
the cracks produces large electric field gradients causing electrons
to be emitted from the negatively charged crack face. Testing this idea
would require knowledge of where the electrons originate. Drs. Tom
Reddyhoff and Julian Le Rouzic from Imperial College, London have
solved this problem by incorporating a microchannel plate electron
detector in their tribometer [2]. Microchannel plates form the basis of
image intensifiers, and consist of an insulating plate containing
thousands of small holes, much like a sieve. A high voltage is placed
across the faces of a plate, so that when an electron enters one of the
holes and strikes the wall of the hole, it can eject several electrons. The
high voltage causes the electrons to be accelerated so that when they hit
the wall again, they eject even more electrons. The stream of electrons
that exits the holes after this process has been repeated many times is
imaged by a fluorescent screen, as in an old-style television, and
produces a much-amplified image of the emitted electron distribution.
Reddyhoff and Le Rouzic placed their electron detector facing their

tribometer, which was installed a vacuum chamber to avoid gas effects (see Fig. 1).

The researchers carried out their experiments using a 100 μm diameter diamond tip sliding at 44 mm/s under a load of 0.4 N on a rotating aluminum disk that had been anodized to form a 5 μm thick oxide film. The electron distributions were measured with a fast camera.

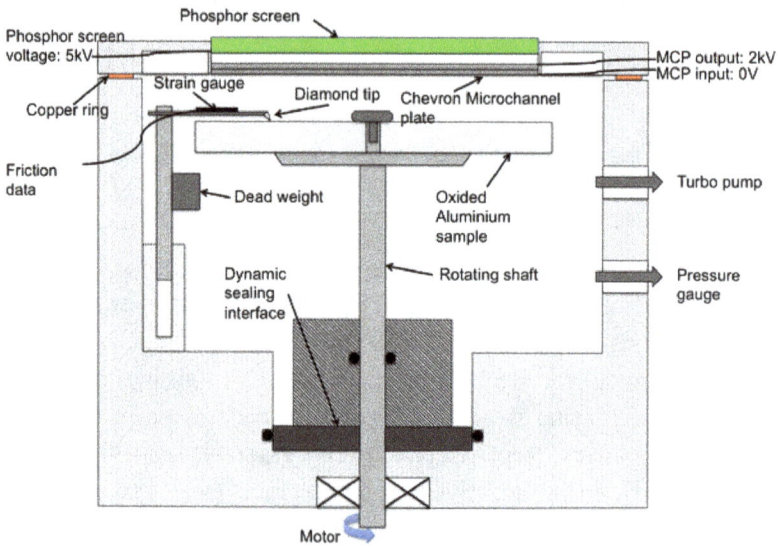

Fig. 1 Schematic diagram of the apparatus. By kind permission of Springer Science+Business Media from Reference [2].

They first carried out a statistical analysis of the total signal produced while rubbing, which revealed that the electron emission was not random — indicating that it must have some deterministic origin. They then examined the electron-emission images as a function of the number of cycles and the angular position on the disk. They found that the largest intensity occurred at the inlet of the contact, suggesting that the electrons were formed by plowing of surface material during abrasion.

However, two other patterns were also observed. The most curious (Fig. 2) showed straight lines occurring repeatedly at specific disk locations and emanating from the contact at an angle of around 60° to the direction of the wear track. It was suggested that these distributions are caused by cracks formed in or along the wear track, in agreement with the idea that electron emission can occur by charge separation during crack formation.

In the third pattern, an initial burst of electrons from the contact was followed by an emission pattern that rotated with the disk. This emission persisted for more than 1 second after exiting the contact. This long time scale is unlikely to be due to propagating cracks and may be due to surface charging.

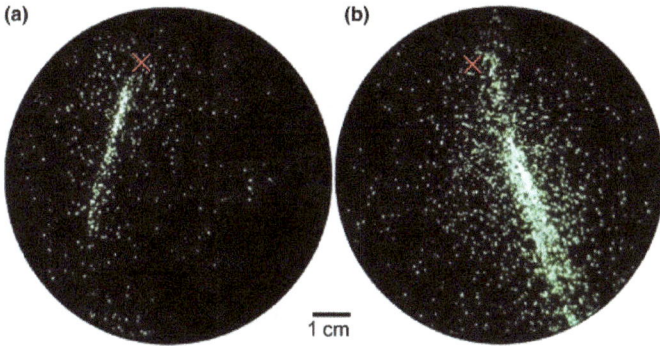

Fig. 2 Spatial distribution of tribo-emitted electrons. By kind permission of Springer Science+Business Media from Reference [2].

Being able to resolve triboelectron emission in space and time has revealed the complexity of the process but, at this point, the explanations for the different emission patterns are speculative. However, identifying the positions at which the emission occurs will allow the group to examine those positions using surface-analytical techniques to test the speculations. The authors also surmise that imaging electron emission could provide an *in situ* approach to monitoring crack formation and growth.

Tribology and Lubrication Technology
October 2014, 70(10) p112

Further Reading:

[1] Tysoe. W.T. and Spencer, N.D. (2009). X-rays by Triboluminescence, Tribology and Lubrication Technology, 65, pp. 72.

[2] Le Rouzic, J. and Reddyhoff, T. (2014). Spatially Resolved Triboemission Measurements. Tribology Letters, 55, pp. 245–252.

Index